Mathematics:
Analysis and Approaches HL
FOR THE IB DIPLOMA

Ian Lucas

PEAK

Published by:
Peak Study Resources Ltd
1 & 3 Kings Meadow
Oxford OX2 0DP
UK

www.peakib.com

Mathematics: Analysis and Approaches HL
Study & Revision Guide for the IB Diploma

ISBN 978-1-913433-01-7

© Ian Lucas 2020

Ian Lucas has asserted his right under the Copyright, Design and Patents Act 1988 to be identified as the author of this work.

All rights reserved. No part of this publication may be reproduced, stored in a retrieval system, or transmitted in any form or by any means, without the prior permission of the publishers.

PHOTOCOPYING ANY PAGES FROM THIS PUBLICATION, EXCEPT UNDER LICENCE, IS PROHIBITED

Peak Study & Revision Guides for the IB Diploma have been developed independently of the International Baccalaureate Organization (IBO). 'International Baccalaureate' and 'IB' are registered trademarks of the IBO.

Orders: books may be ordered directly through the publisher's website www.peakib.com, or to enquire about local stockists please contact us at books@peakib.com (schools@peakib.com for educational establishments).

Printed and bound in the UK by CPI Group (UK) Ltd, Croydon CR0 4YY

www.cpibooks.co.uk

Contents

About this book   v
Acknowledgements   v
The Non-Calculator Paper   vi

Chapter 1: NUMBER AND ALGEBRA 1

    1.1 Number Systems   1
    1.2 Accuracy and Standard Form   2
    1.3 Sequences and Series   3
    1.4 Sequences and Series: Applications   6
    1.5 Exponents   8
    1.6 Logarithms   11
    1.7 Permutations and Combinations   13
    1.8 The Binomial Expansion   15
    1.9 Partial Fractions   18
    1.10 Complex Numbers   20
    1.11 The Complex Plane   22
    1.12 De Moivre's Theorem   24
    1.13 Deductive proof   27
    1.14 Proof by Induction   29
    1.15 Proof by Contradiction   31
    1.16 Systems of Equations   31
    Number and Algebra: Long Answer Questions   34

Chapter 2: FUNCTIONS 37

    2.1 Basics of Functions   37
    2.2 Graphs of Functions   42
    2.3 Linear Functions   47
    2.4 Reciprocal Functions   48
    2.5 Quadratic Functions   51
    2.6 Solving Quadratic Equations   52
    2.7 Polynomial Functions   55
    2.8 Sum and Product of Roots   57
    2.9 Inequalities   59
    2.10 Exponential and Logarithmic Functions   60
    Functions: Long Answer Questions   62

Chapter 3: GEOMETRY AND TRIGONOMETRY 66

    3.1 Solution of Triangles   66
    3.2 3-D Geometry   71
    3.3 Cylinder, Sphere and Cone   73

3.4 Circular Functions   75
3.5 Trigonometric Identities   77
3.6 Solving Trigonometric Equations   79
3.7 Graphing Periodic Functions   81
3.8 Basics of Vectors   82
3.9 Scalar (Dot) Product   84
3.10 Vector (Cross) Product   86
3.11 Equations of Lines   87
3.12 Equations of Planes   88
3.13 Intersections   90
3.14 Angles in Three Dimensions   93
Geometry and Trigonometry: Long Answer Questions   95

Chapter 4: STATISTICS AND PROBABILITY 99

4.1 Definitions   99
4.2 Averages   99
4.3 Measures of spread   103
4.4 Correlation   107
4.5 Sampling Methods   110
4.6 Probability Notation and Formulae   112
4.7 Lists and Tables of Outcomes   113
4.8 Venn Diagrams   114
4.9 Tree Diagrams   116
4.10 Discrete Probability Distributions   118
4.11 The Binomial Distribution   120
4.12 The Normal Distribution   124
4.13 Continuous Distributions   128

Chapter 5: CALCULUS 137

5.1 Differentiation – The Basics   137
5.2 Differentiation from First Principles   138
5.3 L'Hôpital's Rule   139
5.4 The Chain Rule   140
5.5 Product and Quotient Rules   142
5.6 Second Derivative   144
5.7 Graphical Behaviour of Functions   146
5.8 Applications of Differentiation   147
5.9 Implicit Differentiation   152
5.10 Indefinite Integrals   154
5.11 Definite Integrals   156
5.12 Applications of Integration   157
5.13 Integration by Substitution   161

    5.14 Integration by Parts   163

    5.15 Integration using Partial Fractions   165

    5.16 General Methods for Integration   166

    5.17 Differential Equations   168

    5.18 Maclaurin Series   172

    Calculus: Long Answer Questions   175

Chapter 6: MAXIMISING YOUR MARKS 179

Chapter 7: PRACTICE QUESTIONS 181

    Answers to Practice Questions   188

About this book

This is a study guide, not a text book. Its aim is to review each part of the subject and ultimately prepare you for your exams. Covering the complete syllabus, I show you how the topics can translate into questions, giving you plenty of tips and shortcuts.

The exam is not so much a direct test of your knowledge and understanding (you will not get a question which begins "What do you know about...?"); but a test of how you use your knowledge and understanding to solve mathematical problems. So the emphasis in this book is on how to answer questions. In particular you will find plenty of fully worked examples in the text (those similar to Paper 1 exam questions are shaded green), as well as further exercises where it is useful to see how questions can probe a topic from different angles. At the end of each chapter you will find longer questions which are similar in style to those in Paper 2 in the exam – by their nature, such questions may need knowledge of several different areas of the syllabus – and these are indicated with a red band. Answers are provided in the book, but for full working you will need to visit the Mathematics area of the Peak Study Resources website at www.peakib.com.

Question (Paper 1 style)	
Worked answer	Hints and tips

Question (Paper 2 style)

Answer

You are expected to be able to understand and use your graphic display calculator (GDC) to the full in many areas of the syllabus. Indeed, some questions require you to use, for example, the graphing or equation solving features. Since different people use different calculators, it is not possible for this book to explain the detail of their use; but I have indicated calculator tips (using the symbol 📱), and also questions which require calculator use. The more you can use your GDC, the more proficient you will become.

If you want more help with using calculators we have written separate guides for the most popular ones that show you techniques and give plenty of practice with worked examples.

This book is just one resource that will help you prepare for your exams; another is the set of short videos on the website which lead you through the working and solutions of a wide range of exam-style questions. This area of the website is updated as we add more videos to the resource bank. I have indicated in the text if a question has a video solution: look for the video symbol ▶.

I have liberally splashed notes boxes in the margins throughout the book. These contain hints, warnings, exam tips, "dos and don'ts", suggestions... do read them all, as well as the blue text in the question boxes. There's so much information which can help you with those precious extra marks. And there are similar yellow boxes which contain links to other pages, websites, videos, blogs.

Notes box – please read what they contain!

Reference box; leads you to useful resources.

Acknowledgements

I am enormously grateful to Peter Gray, an IB examiner currently at Munich International School, who has proof read this book and, in the process, made some eminently sensible suggestions for numerous improvements; he has also tactfully pointed to a number of errors in both the text and the calculations which I have gratefully corrected! Any remaining errors are entirely my responsibility, and I would be very happy to hear from readers who find any further errors, or who have suggestions for improvements.

Through my work with Oxford Study Courses I have been privileged, over the last 20 years, to help many students revise towards their IB Mathematics exams, and much of what I have learnt from teaching them has been distilled into this book. I would value any feedback so that later editions can continue to help students around the world. Please feel free to e-mail feedback@peakib.com, or comment via the normal social media channels.

Ian Lucas

MATHEMATICS: ANALYSIS AND APPROACHES HL

The Non-Calculator Paper

The format of the two exam papers is the same – Section A consisting of short answer questions, and Section B comprising extended response questions. However, calculators are only allowed to be used in Paper 2.

It is not intended that Paper 1 will test your ability to perform complicated calculations with the potential for careless errors. It is more to see if you can analyse problems and provide reasoned solutions without using your calculator as a prop. However, this doesn't mean there will be no arithmetic calculations. You should, for example, be able to:

Add and subtract using decimals and fractions

Examples:

$18.43 + 12.37$; $2\frac{1}{2} + 3\frac{2}{5}$

Multiply using decimals and fractions

And it would be a good idea to brush up on your multiplication tables – you need them all over the place.

Examples:

432×12 ; 12.6×5 ; $\frac{1}{2} \times \frac{2}{5} + \frac{2}{3} \times \frac{1}{4}$; $(2 \times 10^6) \times (5.1 \times 10^{-4})$

Carry out simple divisions using decimals and fractions

Don't forget that divisions can be written as fractions. For example:
$9 \div 15 = \frac{9}{15} = \frac{3}{5} = 0.6$

Examples:

$14 \div 0.02$; $1\frac{1}{2} \div \frac{3}{5}$

Find x as a fraction in its simplest form if $999x = 324$

Fraction simplification can also help with more complex divisions:

eg: Convert 81 km/h to m/s

$$\frac{81 \times 1000}{3600} = \frac{81 \times 10}{36} = \frac{9 \times 10}{4} = \frac{9 \times 5}{2} = \frac{45}{2} = 22.5 \, \text{m/s}$$

Percentage calculations

Examples:

15% of 600kg; Increase 2500 by 12%

What is 150 as a percentage of 500?

Quadratic equations

You will be called on to solve quadratic equations many times in the papers. Solving by factorisation is easier than using the formula when you are not using a calculator.

Examples:

Solve $x^2 + 7x - 60 = 0$; $3x^2 - 19x + 20 = 0$

Answers in exact form:

When a question asks for the answer to be in "an exact form" this means it must not be given as a rounded decimal. Typically, the answer will contain one (or more) of the following: square roots, logs, π, e.

Chapter 1: NUMBER AND ALGEBRA

1.1 Number Systems

Different situations require different types of number. For example, populations of countries will always be given as positive whole numbers, whereas the division of a reward will require the use of fractions. These are known as *number systems*, and the ones you need to know are:

- *Natural numbers* – positive whole numbers.
- *Integers* – whole numbers including negatives and zero.
- *Rationals* – numbers which can be written as fractions.
- *Irrationals* – numbers which can't be written as fractions.
- *Reals* – the rationals and the irrationals put together. The reals will include every possible number you could meet in the course.

> You need to know the conventional symbols used for the main number systems:
> $\mathbb{N}$ = Natural numbers
> $\mathbb{Z}$ = Integers
> $\mathbb{Q}$ = Rational numbers
> $\mathbb{R}$ = Real numbers
> $\mathbb{Z}^+$ = Positive integers
> $\mathbb{C}$ = Complex numbers

The diagram below shows how the sets are related to each other. For example, every integer can be written as a rational (such as $4 = \frac{4}{1}$), so integers are a subset of rationals.

Reals = Rationals ∪ Irrationals

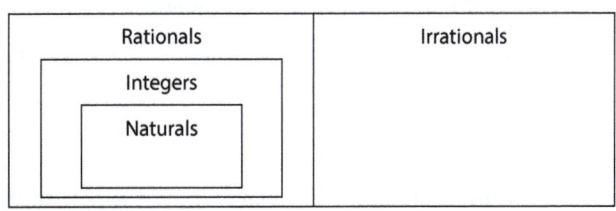

Decimals do not seem to feature in the list above – are they rational or irrational?

- *Recurring decimals* can always be written as fractions so they are rational numbers.
- *Terminating decimals* can also be written as fractions, so they are rational numbers too.
- *Non-recurring, non-terminating decimals* (ie they carry on for ever and never repeat) can never be written as an exact fraction, so they are irrational numbers.

Exact values: $\sqrt{4} = 2$ since 4 is a square number. However, $\sqrt{10}$ cannot be written exactly, like the majority of square roots. It is 3.16228… (the dots indicating that the decimal places continue without recurring). To 4 significant figures $\sqrt{10} = 3.162$, but what do you do if the questions asks you to give an *exact* value? The answer is to use square root notation:

$$x^2 = 10 \Rightarrow x = \sqrt{10}$$

and this (or other equivalent surds) is the only exact way to write down the solution. And, especially if this is an intermediate answer to a question, it is usually better for calculation purposes.

MATHEMATICS: ANALYSIS AND APPROACHES HL

Example: Find the lengths of a and b in the diagram, giving your answers in an exact form.

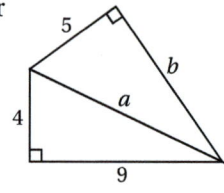

Solution: $a^2 = 9^2 + 4^2 = 97 \Rightarrow a = \sqrt{97}$

$b^2 = a^2 - 5^2 = 97 - 25 = 72$

So $b = \sqrt{72}$

The calculation would have been longer (and possibly less accurate) if we had worked out $\sqrt{97}$ as a decimal and used that.

1.2 Accuracy and Standard Form

When answering questions which have numerical solutions, it is important to understand how to round to an appropriate level of accuracy. And for very large or very small numbers, it is necessary to use standard form.

Accuracy: If there are 6 people in a room, then 6 is perfectly accurate. However, a length given as 6 cm implies that it lies between 5.5 cm and 6.5 cm, and these values are known as the *lower* and *upper bounds*. In theory the upper bound of 6 cm is $6.4\dot{9}$ (ie 6.49999999...), so we should be writing:

$5.5 \leq 6 < 6.5$ (*note the two different symbols*)

It's also important to realise that the number 6.0 implies a greater accuracy:

$5.95 \leq 6.0 < 6.05$

In other words, 6 and 6.0 are different numbers when you consider the range of values each represents.

Questions often ask you to answer to a particular number of *significant figures* or *decimal places*. The first significant figure is the first non-zero digit. The first decimal place is the first figure after the decimal point. Try these:

Write the following numbers to 3SF: 41.26, 2096, 21.04, 699.8

Write the following numbers to 1DP: 12.392, 0.061, 4.952

Answers
41.3, 2100, 21.0, 700
12.4, 0.1, 5.0

If a question asks you to answer to a "suitable" degree of accuracy that usually means you should not be *too* accurate. For example, if a diagram gives lengths to 2SF, your answer should also be to 2SF.

But when you are working through a calculation, you should *not* round off at intermediate stages. Keep full calculator accuracy until you get to the answer, then round.

Standard form: Standard form gives us an alternative way of writing very large and very small numbers without using lots of zeroes. For example:

$43\,000 = 4.3 \times 10\,000 = 4.3 \times 10^4$

$23\,000\,000 = 2.3 \times 10\,000\,000 = 2.3 \times 10^7$

If you're not sure what negative powers mean, look ahead to the section on exponents on page 8.

$0.00056 = \dfrac{5.6}{10\,000} = 5.6 \times \dfrac{1}{10\,000} = 5.6 \times 10^{-4}$

$0.000000109 = \dfrac{1.09}{10\,000\,000} = 1.09 \times 10^{-7}$

Ordinary size numbers can also be written in standard form. For example, $12.5 = 1.25 \times 10^1$. Why would you want to do this? Sometimes calculations involving very big and very small numbers can end up as ordinary size numbers! For example:

$(2.4 \times 10^8) \times (3 \times 10^{-9}) = 7.2 \times 10^{-1} = 0.72$

1. NUMBER AND ALGEBRA

It is important that the first part of the number is between 1 and 10. If you do a calculation and the answer comes out as 12×10^4 this is not in standard form: it must be written as 1.2×10^5.

Some general points about standard form:

- A common mistake is to write eg 4.1×10^3 as 4.1^3
- Make sure you know how to enter numbers in standard form on your GDC, and also how to set the GDC to give answers in standard form. But be careful not to use "calculator notation"; for example, some calculators might display 3×10^{12} as 3E12.
- To add or subtract standard form numbers without a calculator, convert numbers back to ordinary form.
- In exam questions, you may be asked to "give your answer in the form $a \times 10^k$, where $1 \leq a < 10$ and $k \in \mathbb{Z}$" – this is just a formal way of defining standard form.

1.3 Sequences and Series

There are many different types of number sequence. You only need to know about two: the *arithmetic sequence* (or *progression*) (AP) and the *geometric sequence* (GP). In an AP each number is the previous number **plus** a constant. In a GP each number is the previous number **multiplied** by a constant. A *series* is the same as a *sequence* except that the terms are added together: thus a series has a *sum*, whereas a sequence does not.

To answer most sequences and series questions, make sure you are familiar with the formulae below.

First, the notation:

u_1 = the first term of the sequence

n = the number of terms in the sequence

d = the common difference (the number added on in an AP)

r = the common ratio (the multiplier in a GP)

u_n = the value of the nth term

S_n = the sum of the first n terms

S_∞ = the sum to infinity

Try working out the values of d and r in the sequences in the examples box above.

Examples

Arithmetic sequences:
3, 5, 7, 9 ...
1.1, 1.3, 1.5, 1.7 ...
11, 7, 3, −1, −5 ...

Geometric sequences:
1, 3, 9, 27 ...
4, 6, 9, 13.5 ...
12, 6, 3, 1.5, 0.75 ...
2, −6, 18, −54 ...

Answers
d: 2, 0.2, −4
r: 3, 1.5, 0.5, −3

The formulae

For an AP:

The value of the nth term: $u_n = u_1 + (n-1)d$

$d = u_{n+1} - u_n$

The sum of n terms: $S_n = \frac{n}{2}(u_1 + u_n) = \frac{n}{2}(2u_1 + (n-1)d)$

For a GP:

The value of the nth term: $u_n = u_1 r^{n-1}$

$r = \dfrac{u_{n+1}}{u_n}$

The sum formulae always start from the first term. If you wanted to sum, say, the 10th term to the 20th term, you would calculate $S_{20} - S_9$. Think about it!

The sum of n terms: $\quad S_n = \dfrac{u_1(r^n - 1)}{r - 1} \quad$ or $\quad \dfrac{u_1(1 - r^n)}{1 - r}$

And for GPs only there is a formula for "the sum to infinity." If the common ratio is in the range $-1 < r < 1$ then the terms get ever smaller and approach zero. In this case, the *sum* of the series will converge on a particular value. To calculate this value:

The sum to infinity: $\quad S_\infty = \dfrac{u_1}{1 - r}$

Series questions often involve algebra as well as numbers. Note that to find d given two consecutive terms in an AP, subtract the first from the second; and to find r in a GP, divide the second by the first.

Example: A geometric series has first two terms 2 and k. What range of values of k will ensure the series converges?

Solution: The common ratio must be between -1 and 1. The common ratio is $\dfrac{k}{2}$ so $-1 < \dfrac{k}{2} < 1 \Rightarrow -2 < k < 2$.

An arithmetic sequence has third term 12 and sixth term 18.

(a) Find the common difference.
(b) Find the first term.
(c) Find the sum of the first 11 terms.

Using the formula for nth term, we get:

$$u_3 = u_1 + 2d = 12$$
$$u_6 = u_1 + 5d = 18$$

(a) Solving simultaneously, $3d = 6$ giving $d = 2$.
(b) Substituting d into the first equation,
$$u_1 + 2 \times 2 = 12 \text{ giving } u_1 = 8.$$
(c) $S_{11} = \dfrac{11}{2}(2 \times 8 + (11 - 1) \times 2)$
$= \dfrac{11}{2}(16 + 20) = 11 \times 18 = 198$

Once we know the first term and the common difference, we can solve any further questions about the sequence.

You will sometimes find that questions use algebra to define a sequence or its terms. Here's an example where you are given successive *sums* of an AP and have to find the terms.

The trick here is that u_1 is the same as S_1, $u_2 = S_2 - S_1$ and so on.

Example: Successive sums of the terms of an arithmetic sequence are given by:
$S_1 = 1 + t$, $S_2 = 3 + 4t$, $S_3 = 6 + 9t$

Find the first two terms, common difference, and an expression for u_n.

Solution: $u_1 = 1 + t$, $u_2 = (3 + 4t) - (1 + t) = 2 + 3t$

$d = (2 + 3t) - (1 + t) = 1 + 2t$

Using the formula, $u_n = u_1 + (n - 1)d = (1 + t) + (n - 1)(1 + 2t)$

You can easily check the answer by substituting $n = 1$ and $n = 2$.

Multiply out and simplify to get $u_n = n + (2n - 1)t$

It often happens that, where formulae are involved, you are given the value of the variable on the left-hand side, and have to find the value of one of the other variables. This will always involve solving an equation, and the sequence and series formulae are no different in this respect from any others.

> (a) Find the number of terms in the geometric series $1 + 3 + 9 + 27 + ... + 59\,049$
> (b) Calculate the sum of the series in part (a).
>
> a) $u_n = u_1 r^{n-1}$
> $59\,049 = 1 \times 3^{n-1}$
> $n = 11$
>
> b) $S_n = \dfrac{u_1(r^n - 1)}{r - 1}$
> $S_{11} = \dfrac{1(3^{11} - 1)}{3 - 1} = 88\,573$
>
> (a) We want to know the value of n for which $u_n = 59\,049$. We know the first term, the common ratio is clearly 3, so we just substitute these values into the relevant formula. It's good practice to write down any formula you're going to use – apart from anything else, it helps you to substitute values accurately.
>
> We can solve this equation using logs, trial and error, or using the solve function on the GDC.

Remember that the common difference and the common ratio can also be negative:

AP with $u_1 = 7$ and $d = -2$: 7, 5, 3, 1, –1, –3 ...

GP with $u_1 = 2$ and $r = -1.5$: 2, –3, 4.5, –6.75 ...

Sequences and series questions often result in simultaneous equations, the unknowns normally being the first term and common difference or ratio.

> The sum of the first, second and third terms of a geometric sequence is 19 and the sum to infinity is 27. Find the common ratio and the first term.
>
> $r = \dfrac{2}{3}, u_1 = 9$

In a similar vein, try working through the following example:

Example: The ratio of the fourth term to the tenth term of an arithmetic sequence is 5:11. The sum of the first term and the third term is 18. Find the sum of the first 50 terms.

Solution: The way to deal with the first part is to recognise that 11 times the fourth term equals 5 times the tenth term. That gives one equation connecting u_1 and d. The second part gives a second equation. Solve simultaneously to find the values of u_1 and d and then use them to find the sum which is 3975.

> Full working can be found on the website.

Ways of defining sequences and series: Other than listing the numbers, there are two methods for defining a sequence:

- *Function definition,* eg $u_n = 2n^2 + 1$, giving 3, 9, 19, 33, ...
- *Recursive definition,* eg $u_1 = 5, u_{n+1} = 1 + \dfrac{1}{u_n}$ giving $\dfrac{6}{5}, \dfrac{11}{6}, \dfrac{17}{11}, ...$

> Note that neither of these sequences is an AP or a GP. *Any* set of numbers forms a sequence.

Sigma notation can be used to turn the function definition of a sequence into a series; in other words, where the terms of the sequence are added together. In the following example, the sigma symbol means "the sum of."

$$\sum_{1}^{4} (k^2 - 2) = (1^2 - 2) + (2^2 - 2) + (3^2 - 2) + (4^2 - 2) = 22$$

MATHEMATICS: ANALYSIS AND APPROACHES HL

Example: Given that $\sum_{k=1}^{k=n}(4k-1)=1176$, find the value of n.

Solution: The series begins $3+7+11+15$, so we are dealing with an AP with $u_1 = 3$ and $d = 4$.

Using the sum formula, $1176 = \frac{n}{2}(6+(n-1)4)$. This multiplies out to give a quadratic equation with solution $n = 24$.

> Full working can be found on the website

Recurring decimals: We can use the sum of a geometric series to find the exact fractional value of a recurring decimal. For example, to write 2.363636... as a fraction:

$$2.363636... = 2 + \frac{36}{100} + \frac{36}{10000} + ...$$

$$= 2 + \frac{\frac{36}{100}}{1 - \frac{1}{100}} \text{ (using the sum to infinity formula)}$$

$$= 2 + \frac{36}{100} \times \frac{100}{99}$$

$$= 2\frac{4}{11}$$

> Can you see why the denominators of the two smaller fractions on the second line will always be the same?

Sequences and Series: Practice Exercise

1. A geometric sequence has $u_4 = 6$ and $u_8 = 96$. Find the possible values of the common ratio and the corresponding values of the first term.

2. Find the sum to infinity of a geometric series having a second term of -9 and a fifth term of $\frac{1}{3}$.

3. The sum to 20 terms of an arithmetic series is 25 times the first term. If the common difference is d, find the sum to 30 terms in terms of d.

4. Find the sum of the series defined by $\sum_{i=1}^{i=70}(2i-1)$

5. Write the recurring decimal 2.420420420... as a fraction.

6. A circular disc is cut into 12 sectors whose angles are in an arithmetic sequence. The angle of the largest sector is twice the angle of the smallest sector. Find the size of the angle of the smallest sector.

> **Answers**
> 1. 2, −2; 0.75, −0.75
> 2. 20.25
> 3. $1575d$
> 4. 4900
> 5. $2\frac{140}{333}$
> 6. 20°

1.4 Sequences and Series: Applications

One important application of sequences and series is their use in solving financial problems involving interest. If a sum of money is invested, the interest is the amount (expressed as a %) that it earns during each period (usually, but not necessarily, a year).

It is important to remember the single calculations which will:

- Find a percentage of an amount
- Increase or decrease an amount by a percentage

    5 % of 650 = 0.05 × 650

    12 % of 2000 = 0.12 × 2000

    Increase 120 by 5 % = 120 × 1.05

    Increase 4500 by 12 % = 4500 × 1.12

    Decrease 55 by 2 % = 55 × 0.98

1. NUMBER AND ALGEBRA

Simple interest: the interest earned is not added to the total amount which thus stays constant.

- $1000 at 5% simple interest per year will earn $50/year.
  In 10 years, the investment is worth $1000 + 10 \times 50 = \$1500$.

Compound interest: the interest earned is added to the amount invested. Thus the investment grows by a larger amount each year.

- $1000 at 5% compound interest will multiply by 1.05 each year.
  After 1 year, the investment is worth $1000 \times 1.05 = \$1050$
  After 2 years, the investment is worth $1050 \times 1.05^2 = \$1102.50$
  After n years, the investment is worth 1000×1.05^n

Note that with simple interest, the value of the investment is increased by $50 each year and will form an AP. With compound interest, the value will multiply by 1.05 each year and will form a GP.

*Beware of questions where extra money is added to the investment each year **as well as** the interest.*

John invests $1000 at 6% interest, compounded annually, for 16 years. Giving answers to the nearest dollar:

(a) How much is the investment worth after 16 years?

(b) If John removes one quarter of the amount after 4 years, and leaves the remainder to accumulate at 6% until the end of the 16th year, find the final value.

(c) Show if John removes one quarter of the amount at *any* time, and leaves the rest to accumulate at 6%, he will always end up with the same amount as in (b).

(a) $1000 \times 1.06^{16} = 2540.35$
    $= \$2540$

(b) After 4 years, $1000 \times 1.06^4 = 1262.48$
    Remove one quarter $= 946.86$
    Invest for another 12 years:
    $946.86 \times 1.06^{12} = 1905.26$
    $= \$1905$

(c) After n years, 1000×1.06^n
    Remove one quarter: $= 0.75 \times 1000 \times 1.06^n$
    $= 750 \times 1.06^n$
    Invest for $(16 - n)$ years:
    $750 \times 1.06^n \times 1.06^{16-n} = 750 \times 1.06^{16}$
    $= 1905.26$

The first two parts of this question are straightforward applications using compound interest. Note that, although rounding answers to the nearest dollar, I have used the full calculator answer to part (a) when working out part (b).

At first sight part (c) looks much trickier, but is in fact exactly the same as part (b), using n instead of 4.

The key to the algebraic simplification is to use the rule of powers to simplify 1.06^{16-n}.

There is also a clue in the question: if we are to get the same result whatever the value of n, then n must cancel out at some stage leaving a constant.

Sometimes a sneaky question uses a time period other than a year, but the calculations are done in the same way. If an investment attracts compound interest of 6% per annum, paid every month, then we can regard this as being the same as 0.5% per month. To find the value of an investment after 6 years, the multiplier would be 1.005^{72} (72 months at 0.5% per month).

Depreciation: Investments attracting interest appreciate in value, whereas some items will lose value. For example, if a car was bought new for €13000 and then depreciated by 30% each year, after three years it would be worth $13000 \times 0.7^3 = €4459$. Radioactive decay provides another example of a decreasing quantity.

Example: A radioactive substance decays such that it loses 4% of its mass per year. After how many full years would 10 kg first reduce to less than 5 kg?

Solution: We need to solve $10 \times 0.96^t < 5$, giving $t > 16.979...$ so $t = 17$ years.

> See the website for more detail about solving this inequality

The previous example showed a non-financial application. Here's another one. Note that the question doesn't mention sequences and series – you have to be alert to the fact that this is the area of the syllabus it relates to.

> Each day a runner trains for a 9 km race. On the 1st day she runs 1000 m, and then increases the distance by 250 m on each subsequent day.
>
> (a) On which day does she run a distance of 9 km?
>
> (b) What is the total distance in km she will have run in training by the end of that day?

(a) $u_n = u_1 + (n-1)d$

$9000 = 1000 + (n-1) \times 250$

$8000 = 250n - 250$

$\therefore \quad n = 33$ days

(b) $S_n = \frac{n}{2}(u_1 + u_n)$

$S_n = \frac{33}{2}(1000 + 9000) = 165\,000$

She will have run 165 000 m = 165 km

In part (a) we are dealing with the terms of an AP. We know the value of the last term, but need the number of terms. Notice that the question has distances in both km and m.

In part (b) we are looking for the sum of the series. It's easier to use the formula containing u_n (the last term) since we know it is 9000.

Don't forget to convert back to km at the end.

1.5 Exponents

What are exponents? *Exponent* is another word for power or index. You must understand the meaning of negative and fractional powers as well as positive, whole number powers. You must also be very familiar with the rules for using powers. First, let's look at powers of 2:

Powers up here have the conventional meaning of multiplying a number by itself several times.	$2^4 = 16$ $2^3 = 8$ $2^2 = 4$
a^1 is always a for all values of a a^0 is always 1 for all values of a	$2^1 = 2$ $2^0 = 1$
Negative powers ***never*** make the number itself negative. A negative power means "take the reciprocal". So, $a^{-n} = \frac{1}{a^n}$	$2^{-1} = \frac{1}{2} = \frac{1}{2^1}$ $2^{-2} = \frac{1}{4} = \frac{1}{2^2}$ $2^{-3} = \frac{1}{8} = \frac{1}{2^3}$

1. NUMBER AND ALGEBRA

Fractional powers always involve *roots*.

Power	Meaning
$x^{\frac{1}{2}}$	$\sqrt{x}$
$x^{\frac{1}{3}}$	$\sqrt[3]{x}$
$x^{\frac{3}{2}}$	$\sqrt{x^3} = (\sqrt{x})^3$
$x^{-\frac{2}{5}}$	$\dfrac{1}{\sqrt[5]{x^2}}$

Examples:
$2.5^1 = 2.5$
$4^{-2} = \dfrac{1}{16}$
$\left(\dfrac{2}{3}\right)^{-3} = \left(\dfrac{3^3}{2^3}\right) = \dfrac{27}{8}$
$8^{\frac{5}{3}} = \left(\sqrt[3]{8}\right)^5 = 32$

In general, $x^{\frac{1}{n}} = \sqrt[n]{x}$ and $x^{\frac{m}{n}} = \sqrt[n]{x^m} = \left(\sqrt[n]{x}\right)^m$.

Laws of exponents: The rules which follow occur in all sorts of mathematical situations and you should learn them carefully:

Examples

- $a^x \times a^y = a^{x+y}$ $\qquad 2^{x+3} = 2^x \times 2^3 = 8 \times 2^x$
- $a^x \div a^y = a^{x-y}$ $\qquad \dfrac{x^3}{x^5} = x^{3-5} = x^{-2}$ or $\dfrac{1}{x^2}$
- $(a^x)^y = a^{xy}$ $\qquad 9^x = (3^2)^x = 3^{2x}$
- $(ab)^x = a^x b^x$ $\qquad (3x)^3 = 3^3 \times x^3 = 27x^3$

Solving equations containing exponents: First, ensure that you are familiar with common powers of integers from 2 to 6.

$2^2 = 4, 2^3 = 8, 2^4 = 16, 2^5 = 32, 2^6 = 64, 2^7 = 128.$

$3^2 = 9, 3^3 = 27, 3^4 = 81, 3^5 = 243.$

$4^2 = 16, 4^3 = 64, 4^4 = 256.$

$5^2 = 25, 5^3 = 125, 5^4 = 625.$

$6^2 = 36, 6^3 = 216.$

Since negative powers involve reciprocals, it is likely that questions involving negative powers will require you to manipulate fractions.

You may literally need to know these backwards. For example, to solve $243 \times 3^{2x} = 1$ you would need to recognise that $243 = 3^5$ ($x = -2.5$).

Fractional powers always involve roots. In the non-calculator paper, the laws of exponents will help you to simplify expressions. For example, we can use $9 = 3^2$ to simplify $\dfrac{\sqrt{3} \times 9^{\frac{2}{3}}}{\sqrt[3]{9}}$.

$$\dfrac{\sqrt{3} \times 9^{\frac{2}{3}}}{\sqrt[3]{9}} = \dfrac{3^{\frac{1}{2}} \times (3^2)^{\frac{2}{3}}}{(3^2)^{\frac{1}{3}}} = \dfrac{3^{\frac{1}{2}} \times 3^{\frac{4}{3}}}{3^{\frac{2}{3}}} = \dfrac{3^{\frac{11}{6}}}{3^{\frac{2}{3}}} = 3^{\frac{7}{6}}$$

We can use a similar technique to solve an equation such as $4^x - 4 \times 8^x = 0$. The trick is to look for a common base, in this case 2.

$$(2^2)^x - 2^2 \times (2^3)^x = 0$$
$$2^{2x} = 2^{2+3x}$$
$$2x = 2 + 3x$$
$$x = -2$$

This one's a bit harder: $8^x = 0.25^{3x-1}$. Think of 0.25 as a fraction, and then convert it to a power of 2. The final answer is $x = \dfrac{2}{9}$.

MATHEMATICS: ANALYSIS AND APPROACHES HL

> Solve the equation $9^x = 27^{1-x}$.
>
> | | |
> |---|---|
> | $(3^2)^x = (3^3)^{1-x}$ | As in the previous examples, we need to use a common base. In this case it will be 3. |
> | $3^{2x} = 3^{3-3x}$ | |
> | $2x = 3 - 3x$ | |
> | $5x = 3$ | The answer could also be given as 0.6. Fractions and decimals are different forms of the same number. |
> | $x = \frac{3}{5}$ | |

See page 31 and page 47

The next exam-style question is obviously about exponents but, look carefully, and it's actually a quadratic equation. In other words, it's of the form something^2 + something + number. You will also find this form when dealing with trigonometric equations and equations involving e^x.

> (a) Show that the equation $(2^x)^2 + 2^x - 20 = 0$ can be written in the form $(2^x + a)(2^x + b) = 0$, where a and b are integers to be found.
>
> (b) Hence solve the equation for x and explain why there is only one solution.
>
> | | |
> |---|---|
> | (a) $(2^x)^2 + 2^x - 20 = 0$ | It may help to use y as the base of quadratic, then replace y with 2^x. |
> | $y^2 + y - 20 = 0$ | |
> | $(y+5)(y-4) = 0$ | When solving equations in exams, it's always a good idea to substitute your answer into the original equation to check you are right. In this case, with $x = 2$, we get |
> | $(2^x + 5)(2^x - 4) = 0$ | |
> | (b) So $2^x = -5$ or $2^x = 4$. But exponentials cannot take negative values, so the only solution is $2^x = 4$, giving $x = 2$. | $4^2 + 4 - 20 = 16 + 4 - 20 = 0$ |

Surds: Make sure you are entirely familiar with the rules for manipulating surds. In particular:

$$\sqrt{a} \times \sqrt{b} = \sqrt{ab}, \quad \frac{\sqrt{a}}{\sqrt{b}} = \sqrt{\frac{a}{b}}$$

$$\sqrt{a} + \sqrt{b} \neq \sqrt{a+b}, \quad \sqrt{a} - \sqrt{b} \neq \sqrt{a-b}$$

Thus:
$$\sqrt{2} \times \sqrt{18} = \sqrt{36} = 6$$

$$\sqrt{50} + \sqrt{98} = \sqrt{25}\sqrt{2} + \sqrt{49}\sqrt{2} = 5\sqrt{2} + 7\sqrt{2} = 12\sqrt{2}$$

$$\sqrt{\frac{49}{100}} = \frac{\sqrt{49}}{\sqrt{100}} = \frac{7}{10}$$

It is sometimes useful when manipulating surds to think of the rules of algebra - in the second example above, compare $5\sqrt{2} + 7\sqrt{2}$ with $5x + 7x$. This should enable you to quickly calculate the values of $(4\sqrt{3})^2$ and $(2 + \sqrt{5})(2 - \sqrt{5})$. (Answers: 48 and −1)

You should also be able to *rationalise* the denominators of fractions which contain surds. If the denominator is of the form $\sqrt{a}$, then multiply numerator and denominator by $\sqrt{a}$; if the denominator is of the form $a + \sqrt{b}$, then multiply numerator and denominator by $a - \sqrt{b}$. In both cases, the denominator will become a rational number. (The process is very similar to the division of complex numbers.)

For example: $\quad \dfrac{4}{3-\sqrt{3}} = \dfrac{4}{3-\sqrt{3}} \times \dfrac{3+\sqrt{3}}{3+\sqrt{3}} = \dfrac{12 + 4\sqrt{3}}{9 - 3} = \dfrac{12 + 4\sqrt{3}}{6} = 2 + \dfrac{2\sqrt{3}}{3}$

1. NUMBER AND ALGEBRA

1.6 Logarithms

What is a logarithm? The mapping diagram in the notes box shows the function $f(x) = 2^x$ applied to a few integers. The inverse of this function would map $8 \to 3$, $4 \to 2$ and so on: in other words, it would find what power of 2 gives the required number. As shown at the bottom of the diagram, the inverse is the logarithm function. So the logarithm to the *base 2* of a number is the power of 2 which gives the number. For example, $\log_2 16 = 4$. It may be helpful to think of the relationship of the log(arithm) function to the power function as similar to that between the square root function and the square function.

> $x \to 2^x$
> $3 \to 8$
> $2 \to 4$
> $1 \to 2$
> $0 \to 1$
> $-1 \to 0.5$
> and $\log_2 x \leftarrow x$

Change of base: Logarithms can be to any base, and occasionally you may need to change from one base to another, in which case use the change of base formula:

$$\log_b a = \frac{\log_c a}{\log_c b}$$

For example: simplify $\log_a 18 \times \log_3 a = \dfrac{\log_3 18}{\log_3 a} \times \log_3 a = \log_3 18$

> **Examples:**
> $\log_3 27 = 3$
> $\log_{10} 0.1 = -1$
> $\log_2(\sqrt{2}) = \frac{1}{2}$
> $\log_5(5^x) = x$

Laws of logarithms: Because logarithms are just powers, the laws of logarithms are very similar to the laws of exponents. You should be very familiar with them, although you will find them in the formula book. These rules apply to logs with any base.

- $\log a + \log b = \log(ab)$
- $\log a - \log b = \log(a/b)$
- $\log a^n = n \log a$

The last gives a useful method for solving equations with powers in because it can be used to "bring the power down."

Example: Solve $2^x = 13$

The logs can be to any base, so in practice use the log key on your calculator (ie base 10).

$$2^x = 13$$
$$\log(2^x) = \log 13$$
$$x \log 2 = \log 13$$
$$x = \frac{\log 13}{\log 2} = 3.70$$

> A common mistake is to write $\log ax^2$ as $2\log ax$. This would only work if the square applied to the a as well. But we could also use the first law to get:
> $\log ax^2 = \log a + \log x^2$
> $= \log a + 2\log x$

It's almost certain that you will get exam questions which involve the laws of logarithms. The following example is typical.

MATHEMATICS: ANALYSIS AND APPROACHES HL

> If $\log_a 2 = x$ and $\log_a 5 = y$, find in terms of x and y expressions for (a) $\log_2 5$ and (b) $\log_a 20$.

(a) $\log_b a = \dfrac{\log_c a}{\log_c b}$ So $\log_2 5 = \dfrac{\log_a 5}{\log_a 2}$ $= \dfrac{y}{x}$ (b) $\quad 20 = 5 \times 2^2$ So $\log_a 20 = \log_a (2^2 \times 5)$ $\quad = \log_a 2^2 + \log_a 5$ $\quad = 2\log_a 2 + \log_a 5$ $\quad = 2x + y$	We are given logs to the base a and, in part (a), we want a log to the base 2. It follows that we need to use the change of base formula. Don't get confused by the letters. The "a" in the initial formula is not the same "a" as in the question! In part (b) we need to rewrite 20 in terms of 2 and 5, without using addition or subtraction. We then use law 3 followed by law 1.

Logarithms can crop up in all sorts of unlikely places. Examiners quite like including them in sequences and series questions.

Example: The first two terms of an infinite geometric series are $3\log_2 x$ and $\log_2 x$. Find the sum of the series.

Solution: We need the common ratio which is $\dfrac{u_2}{u_1} = \dfrac{\log_2 x}{3\log_2 x} = \dfrac{1}{3}$. Now we know it's an infinite sequence so the sum is $\dfrac{u_1}{1-r} = \dfrac{3\log_2 x}{1 - \frac{1}{3}} = \dfrac{9\log_2 x}{2}$.

No need to use the laws of logarithms there. But how about this one?

Example: The first two terms of an arithmetic sequence are $\log_2 x$ and $\log_2\left(\dfrac{x}{4}\right)$. Find the common difference, giving your answer as an integer.

Solution: $d = u_2 - u_1 = \log_2\left(\dfrac{x}{4}\right) - \log_2 x = \log_2\left(\dfrac{x/4}{x}\right) = \log_2 \dfrac{1}{4} = -2$

Solving equations containing logarithms: You'll find that equations involving logs reduce to two main types:

$$\log_a(f(x)) = \log_a(g(x)) \quad \text{or} \quad \log_a(f(x)) = b$$

eg: $\log_2 x + \log_2 5 = \log_2(x+1) \Rightarrow 5x = x+1 \Rightarrow x = \dfrac{1}{4}$

> ▶ See video online for the solution of $\log_6 x + \log_6 (x-5) = 2, \; x > 0$

eg: $\log_3 x - \log_3 6 = 2 \Rightarrow \log_3\left(\dfrac{x}{6}\right) = 2 \Rightarrow \dfrac{x}{6} = 3^2 \Rightarrow x = 54$

I can't emphasise too much how important it is to know the laws of exponents and logarithms by heart so that you can easily apply them in a variety of mathematical situations. Also, be absolutely clear what a logarithm is, and understand that $\log_b x = y$ is the same as $x = b^y$.

e^x and $\ln x$: Functions such as 2^x and 3^x are exponential functions; e^x is called "*the exponential function*" because of its importance in modelling functions such as population growth. Like π, e is an irrational number and has an approximate value of 2.718... Its inverse is $\log_e x$ which is normally written as $\ln x$.

12

1. NUMBER AND ALGEBRA

Note that $e^{\ln x} = x$ and $\ln e^x = x$. Use this when simplifying expressions, solving equations or changing the subject of the formula. This rule is true for logs to any base.

Example: If $p = 3 + 2e^{0.04t}$, find t when $p = 7.63$.

Solution:
$$7.63 = 3 + 2e^{0.04t}$$
$$4.63 = 2e^{0.04t}$$
$$2.315 = e^{0.04t}$$
$$\ln 2.315 = 0.04t$$
$$t = \frac{\ln 2.315}{0.04}$$
$$t = 21.0 \text{ (to 3SF)}$$

I've chosen the next exam-style question to illustrate how it's possible to get too hung up on a particular method, in this case assuming the equation will be solved using laws of logarithms.

Solve $2(\ln x)^2 = 3\ln x - 1$	
$2(\ln x)^2 - 3\ln x + 1 = 0$	It is worth remembering that there is no way of simplifying $(\ln x)^2$. It is NOT $2\ln x$.
$(2\ln x - 1)(\ln x - 1) = 0$	But with a term in $(\ln x)^2$, a term in $\ln x$ and a constant term, what have we got?
$\ln x = \tfrac{1}{2}$ or $\ln x = 1$	A quadratic equation...
$x = e^{\tfrac{1}{2}}$ or $x = e$	

Logarithms: Practice Exercise

Try the following questions if you require more practice in the basic manipulation of logarithms. Only number 7 needs a GDC.

1. Solve $\log_5 x + \log_5 3 = \log_5 12$
2. Solve $\log_4 p = 3$
3. Rewrite $\log_{10} 3 + 2\log_{10} 5$ as a single logarithm
4. Find the exact value of x (in terms of a) in the equation $\log_a(2x - 1) = 3$
5. If $\log_3 a = m$ and $\log_3 b = n$, write $\log_3\left(\frac{\sqrt{a}}{b}\right)$ in terms of m and n
6. Let $\log_b 2 = p$ and $\log_b 5 = q$. Find an expression in terms of p and q for:
   (i) $\log_b 10$
   (ii) $\log_b 50$
7. Solve $\log_9 4 + \log_3 x = 3$
8. Find the value of b if $\log_b 7 = \tfrac{1}{2}$
9. Use logarithms to solve $4^{2x+1} = 100$

Answers
1. 4
2. 64
3. $\log_{10} 75$
4. $\frac{a^3 + 1}{2}$
5. $\tfrac{1}{2}m - n$
6. (i) $p + q$
   (ii) $p + 2q$
7. 1.161...
8. 49
9. 1.16 to 3SF

Question 4 is a more complicated form of question 2.

In question 5, first use law 2 then law 3.

In question 7, use the change of base formula to rewrite the first log using logs to base 3.

1.7 Permutations and Combinations

Use *combinations* when you want to select a certain number of objects from a group, but order is unimportant. Use *permutations* when you want to both select and arrange the objects.

13

MATHEMATICS: ANALYSIS AND APPROACHES HL

The multiplication principle: If there are 2 ways of choosing an option A and 3 ways of choosing an option B, then there will be 2 × 3 = 6 ways of choosing option A followed by B. For example, when you throw a die then toss a coin there will be 6 × 2 = 12 possible outcomes.

Combinations: The combinatorial formula $^nC_r = \dfrac{n!}{r!(n-r)!}$ will calculate the number of ways of selecting r objects from n. Thus the number of ways of selecting 3 objects from 5 is $^5C_3 = \dfrac{5!}{3!(5-3)!} = \dfrac{5!}{3!2!} = \dfrac{120}{6 \times 2} = 10$

With larger numbers it is helpful to write the factorials out so that you can see what cancels top and bottom. For example:

$$^7C_4 = \frac{7!}{4!3!} = \frac{7 \times 6 \times 5 \times 4 \times 3 \times 2 \times 1}{(4 \times 3 \times 2 \times 1)(3 \times 2 \times 1)} = \frac{7 \times 6 \times 5}{3 \times 2 \times 1} = 7 \times 5 = 35$$

The cancelling is important because otherwise you would find yourself trying to calculate 7! – not advisable when you're doing an exam! So you will *always* be able to cancel the larger factorial on the bottom with the right hand part of the factorial on the top. In the example above, I then calculated the bottom line as 6, and cancelled this with the 6 in the top line.

> So in this example, I have cancelled the 7! on the bottom with most of the 9! on the top.

With a bit of practice, you shouldn't have to write out the factorials at all.

$$^9C_2 = \frac{9 \times 8}{2 \times 1} = 36$$

Without a calculator, work out the values of:

$^6C_3, {}^9C_1, {}^8C_5, {}^4C_2$ (20, 9, 56, 6)

It is also useful to remember that $^nC_0 = {}^nC_n = 1$, and $^nC_1 = n$.

> We could also turn this into a probability, eg:
>
> *You are choosing a committee of 4 from 13 people. What is the probability that the committee will consist of 2 men and 2 women?*
>
> Answer: $\dfrac{280}{715} = 0.392$

Thus, if you have to choose a committee of 4 from 13 people, the number of possible choices is $^{13}C_4 = 715$ (GDC). What if the 13 consists of 8 men and 5 women, and you want to choose 2 of each? Using the multiplication principle this becomes $^8C_2 \times {}^5C_2 = 280$.

Sometimes constraints are put in:

Example: A football team of 11 players must be chosen from a squad of 15. The captain must be selected, as must one of the two possible goalkeepers. How many possible teams could be chosen?

Solution: In effect, we must select 1 from 1, 1 from 2, then 9 from the remaining 12. Using the combinatorial formula and the multiplication principle, the calculation is $^1C_1 \times {}^2C_1 \times {}^{12}C_9 = 440$

A team of 4 is chosen at random from 5 girls and 6 boys. In how many ways can the team be chosen if there must be more boys than girls?	
$^6C_3 \times {}^5C_1 + {}^6C_4 \times {}^5C_0 = 115$	3 boys, 1 girl or 4 boys, 0 girls

Permutations: There are $n!$ ways of arranging n objects. But if just some of the n objects are to be selected and then arranged, I use the "boxes" method. For example, in how many ways can you choose a first and a second prize winner from 6 dogs in a dog show?

1. NUMBER AND ALGEBRA

Draw two boxes

6 possibilities for the first box

5 possibilities for the second box

| 6 | 5 |

Thus there are 30 possible arrangements.

The boxes method can be used for a variety of permutation problems. Let's consider making integers using the digits 1, 2, 3, 4, 5:

(a) How many 4 digit integers can be formed if each of the digits can be used more than once?

Five possibilities for each box.

| 5 | 5 | 5 | 5 | = 625

(b) How many 4 digit integers can be formed if each of the digits can only be used once?

Five for the first, four for the second, etc.

| 5 | 4 | 3 | 2 | = 120

(c) How many integers between 1000 and 3999 can be formed if each of the digits can only be used once?

*The first digit must be 1, 2 or 3, so three choices.
No restrictions after that.*

| 3 | 4 | 3 | 2 | = 72

(d) How many four digit even numbers can be formed if each of the digits can only be used once?

*The last digit must be 2 or 4, so two choices.
Then no restriction on the first three.*

| 4 | 3 | 2 | 2 | = 48

An interesting variation is a question such as: "How can you arrange 4 green and 2 red books on a shelf such that the red books are not together?" One way of doing this is to list all the ways the two red books can be placed – no need to do the greens as well since they must occupy all the other slots. You should find 10 possibilities, with probability 0.0139.

1.8 The Binomial Expansion

Calculation of binomial coefficients: It is helpful to remember the first few rows of Pascal's Triangle; the blue numbers in the table are called "binomial coefficients." For example, the 5th row, 3rd column is $^5C_3 = 10$.

You should also be able to use your GDC to calculate a single binomial coefficient (using the ^nC_r formula); and a set of binomial coefficients, equivalent to a single row in the table above. There are a couple of ways this can be achieved:

	0th	1st	2nd	3rd	4th	5th
1st	1	1				
2nd	1	2	1			
3rd	1	3	3	1		
4th	1	4	6	4	1	
5th	1	5	10	10	5	1

Using a table: To return, say, the 6th row, set up the function $6\,^nC_r\,x$ (different calculators will do this in different ways), and then look at the table of values where x starts at 0 and increments in steps of 1. Clearly there is no meaning to the values returned when $x > 6$.

MATHEMATICS: ANALYSIS AND APPROACHES HL

Using a list: To return the 6th row, enter $6{}^nC_r\{0, 1, 2, 3, 4, 5, 6\}$ where the curly brackets indicate a list of vales. This will then return another list containing the binomial coefficients.

The Binomial Theorem: The general formula gives you a quick way of multiplying out brackets of the form $(a + b)^n$ where n is a natural number. It is best illustrated with an example.

To expand $(a + b)^4$ each term will have 3 parts to it: the appropriate Pascal's Triangle number (in this case using row 4), a to a power beginning at 4 and reducing to 0, b to a power beginning at 0 and increasing to 4. The 1s that result in the first and last terms reduce those terms to a^4 and b^4.

$$(a + b)^4 = a^4 + 4a^3b + 6a^2b^2 + 4ab^3 + b^4$$

This general form can now be used to expand more specific expressions, for example $(2 - 3x)^4$. When doing this, note:

- Always write out the general form first, then substitute underneath (in this case, $a = 2, b = -3x$).
- Use brackets throughout to ensure correct calculation.
- Use one line to substitute, the next (possibly more than one) to calculate.

$$(2 - 3x)^4 = 2^4 + 4(2)^3(-3x) + 6(2)^2(-3x)^2 + 4(2)(-3x)^3 + (-3x)^4$$
$$= 16 - 96x + 216x^2 - 216x^3 + 81x^4$$

Be careful when substituting a negative number - always use brackets.

Use binomial expansion to express $(1 + \sqrt{7})^3$ in the form $p + q\sqrt{7}$, $p, q \in \mathbb{Z}$	
$(a + b)^3 = a^3 + 3a^2b + 3ab^2 + b^3$ $(1 + \sqrt{7})^3 = 1^3 + 3 \times 1^2 \times (\sqrt{7}) + 3 \times 1 \times (\sqrt{7})^2 + (\sqrt{7})^3$ $= 1 + 3\sqrt{7} + 21 + 7\sqrt{7}$ $= 22 + 10\sqrt{7}$	When substituting numbers for letters, it is safer to substitute first with no calculating, then carefully calculate each term.

Finding individual terms: Sometimes you are asked to find a specific term rather than work out the whole expansion. For example, let's find the term in x^3 in the expansion of $(2 - \frac{1}{2}x)^7$. We need to work out the three constituent parts of the term:

- The binomial coefficient is ${}^7C_3 = 35$
- The power of the x part is 3, so this gives $(-\frac{1}{2}x)^3 = -\frac{1}{8}x^3$
- The powers always add to give the overall power, so the third part is $2^4 = 16$

Thus the overall term is $35 \times 16 \times (-\frac{1}{8}x^3) = -70x^3$

A common problem is to find the "constant term"; that is, the term which is just a number. In other words, it is the term which contains x^0. Watch the video solution for the following question which an examiner clearly enjoyed setting!

Given that the term independent of $x = 1584$, find the two possible values of k in the expansion of $(1 + x^3)\left(\frac{2}{x} + kx^2\right)^6$
$k = 2.2$ or -3

1. NUMBER AND ALGEBRA

Binomial Expansion: Practice Exercise

1. Expand and simplify $\left(x - \frac{2}{x}\right)^5$

2. Find the constant term in the expansion of $\left(2x + \frac{1}{x^2}\right)^6$

3. Find the term in x^2 in the expansion of $(1 - 4x)(2 - x)^3$

   (Expand the second bracket. Then calculate just the terms in x^2 which will result when the brackets are multiplied out.)

4. Determine the first three terms in the expansion of $(1 - 2x)^5(1 + x)^7$.

   (Similar to the previous question, but there's a bit more to do. You need to expand both brackets as far as the terms in x^2, and then combine to find all the terms up to the term in x^2.)

5. (a) Expand

   (i) $(1 + x)^3$

   (ii) $(1 - x)^3$

   (b) Hence write $(1 + x)^3 + (1 - x)^3$ in the form $ax^2 + b$.

   (c) Use your answer to part (b) to find the value of k such that $(1 + k)^3 + (1 - k)^3 = 152$, where k is a positive number.

6. Use the binomial expansion to show that, if $z = 2 + 3i$, then $z^3 = -46 + 9i$. Also expand $(2 - 3i)^3$ and hence prove that $(z^3)^* = (z^*)^3$.

7. When $(1 + x)^n$, $n \in \mathbb{N}$, is expanded in ascending powers of x, the coefficient of x^2 is 66. Find the value of n.

   (You will need to use the combinatorial formula to calculate the binomial coefficient.)

Answers

1. $x^5 - 10x^3 + 40x - 80x^{-1} + 80x^{-3} - 32x^{-5}$
2. 240
3. $54x^2$
4. $1 - 3x - 9x^2$
5. (a)(i) $1 + 3x + 3x^2 + x^3$
   (ii) $1 - 3x + 3x^2 - x^3$
   (b) $2 + 6x^2$
   (c) 5
6. $(2 - 3i)^3 = -46 - 9i$
7. 12

Binomial Theorem for $n \in \mathbb{Q}$: When n is a natural number, the expansion of $(a + b)^n$ will give the equivalent algebraic expression as if the brackets had been multiplied out manually. However, if n is negative, or a fraction, this no longer applies: for example, $(2 + x)^{\frac{1}{2}}$ cannot be multiplied out. Instead, the binomial theorem creates an infinite series in x which approximates to the original expression. Each additional term increases the accuracy. The series is given by:

$$(1 + x)^n = 1 + nx + \frac{n(n-1)}{2!}x^2 + \frac{n(n-1)(n-2)}{3!}x^3 + \ldots$$

This is also the Maclaurin series for $(1 + x)^n$. See page 172.

To ensure this works properly, there are a couple of constraints:

- The first term in the bracket must be 1
- $|x| < 1$ (ensuring successive terms decrease in value)

But what if the first term isn't 1? Here are a couple of examples which show what you have to do:

- $(4 + x)^{\frac{1}{2}} = \left(4\left(1 + \frac{x}{4}\right)\right)^{\frac{1}{2}} = 4^{\frac{1}{2}} \times \left(1 + \frac{x}{4}\right)^{\frac{1}{2}} = 2\left(1 + \frac{x}{4}\right)^{\frac{1}{2}}$

- $\frac{1}{3 + x} = \frac{1}{3\left(1 + \frac{x}{3}\right)} = \frac{1}{3}\left(1 + \frac{x}{3}\right)^{-1}$

In the second example, note the 3 which remains in the denominator. It's an easy mistake to simplify as $3\left(1 + \frac{x}{3}\right)^{-1}$

In the first example, the series will work for $\left|\frac{x}{4}\right| < 1 \Rightarrow |x| < 4$, and in the second series $|x| < 3$.

The binomial series for the first example is calculated as follows:

$$(4+x)^{\frac{1}{2}} = 2\left(1+\frac{x}{4}\right)^{\frac{1}{2}} \approx 2\left(1 + \frac{1}{2} \times \frac{x}{4} + \frac{\left(\frac{1}{2}\right)\left(-\frac{1}{2}\right)}{2}\left(\frac{x}{4}\right)^2 + \frac{\left(\frac{1}{2}\right)\left(-\frac{1}{2}\right)\left(-\frac{3}{2}\right)}{6}\left(\frac{x}{4}\right)^3\right)$$

$$= 2\left(1 + \frac{x}{8} - \frac{x^2}{128} + \frac{x^3}{3072}\right)$$

$$= 2 + \frac{x}{4} - \frac{x^2}{64} + \frac{x^3}{1536}$$

Let's see how accurate the approximation is. Suppose we want to calculate $\sqrt{4.1}$ by putting $x = 0.1$. From the calculator, $\sqrt{4.1} = 2.024846$ to 6DP. The binomial series gives us:

2.025 to two terms

2.0248438 to three terms

2.0248444 to four terms

> Use binomial expansion to expand $(1-x)(1+3x)^{-2}$ as far as the term in x^2.
>
> $(1+3x)^{-2} = 1 + (-2)(3x) + \frac{(-2)(-3)}{2!}(3x)^2 + ...$ *Expand the second bracket first, then multiply out the result by the first bracket.*
>
> $\phantom{(1+3x)^{-2}} = 1 - 6x + 27x^2 + ...$
>
> Thus $(1-x)(1+3x)^{-2} = (1-x)(1-6x+27x^2)$
>
> $\phantom{Thus (1-x)(1+3x)^{-2}} = 1 - 6x - x + 6x^2 + 27x^2 - 27x^3 + ...$
>
> $\phantom{Thus (1-x)(1+3x)^{-2}} = 1 - 7x + 33x^2$ as far as the term in x^2

1.9 Partial Fractions

As with numeric fractions, algebraic fractions can be added together to give a single fraction. For example:

$$\frac{2}{x+1} + \frac{3}{x-2} \equiv \frac{2(x-2)+3(x+1)}{(x+1)(x-2)} \equiv \frac{5x-1}{(x+1)(x-2)}$$

The reverse process splits a single fraction into *partial fractions*. Both forms have their uses. A single fraction, for example, is the most helpful when sketching the graph of a function; the partial fraction form can be used to help with integration. You will only need to deal with examples such as the one above; that is, with two linear terms in the denominator, and where the numerator is either a constant or a linear term.

See page 165 for more on integrating using partial fractions.

Substituting x values method: Express $\dfrac{2x-1}{(x-2)(x-3)}$ in partial fractions:

1. Rewrite as two fractions, using constants as denominators.

    $\dfrac{2x-1}{(x-2)(x-3)} \equiv \dfrac{A}{x-2} + \dfrac{B}{x-3}$

2. Multiply both sides by the left hand denominator, and cancel on RHS

    $2x - 1 \equiv A(x-3) + B(x-2)$

3. Substitute values of x which, in turn, make the linear terms 0, and hence evaluate the constants.

    $x = 3$: $5 = 0 + B \Rightarrow B = 5$

    $x = 2$: $3 = -A + 0 \Rightarrow A = -3$

4. Substitute the values of the constants.

    $\dfrac{2x-1}{(x-2)(x-3)} \equiv \dfrac{-3}{x-2} + \dfrac{5}{x-3}$

Why can we just substitute any values of x? Because the two expressions form an identity, and hence will be equal for ***all*** values of x.

1. NUMBER AND ALGEBRA

Cover-up method: This method is quicker, and enables you to write the answers down with very little intermediate working. To find A, you choose the value of x which makes the denominator of the partial fraction containing A equal to zero. In the above example this is $x = 2$. Substitute this value into the left hand side, but "cover up" the same term whilst doing it. This gives $\dfrac{2x-1}{(\ \ )(x-3)} = \dfrac{3}{-1} = -3$. This is the value of A. Now do the same for $x = 3$: $\dfrac{2x-1}{(x-2)(\ \ )} = \dfrac{5}{1} = 5$, and hence $B = 5$.

For a bit of practice, express the following as partial fractions:

$$\dfrac{3x+5}{(x+3)(x-1)}; \quad \dfrac{3-8x}{x(1-x)}; \quad \dfrac{2}{x^2-4x+3}$$

Answers: $\dfrac{1}{x+3} + \dfrac{2}{x-1}; \quad \dfrac{3}{x} - \dfrac{5}{1-x}; \quad \dfrac{1}{x-3} - \dfrac{1}{x-1}$

Alternatively, $\dfrac{-5}{1-x}$ could be written as $\dfrac{5}{x-1}$.

Notice that in the last two answers I have transferred the minus sign from the constant to become a subtraction sign in front of the fraction. Neater.

Partial fractions and binomial series: Partial fractions lead naturally to questions involving binomial series where $n = -1$. Look at the example at the start of the Partial Fractions section. $\dfrac{2}{x+1} + \dfrac{3}{x-2}$ can be rewritten as follows:

$$\dfrac{2}{x+1} + \dfrac{3}{x-2} \equiv \dfrac{2}{1+x} - \dfrac{3}{2-x}$$

$$\equiv \dfrac{2}{1+x} - \dfrac{3}{2\left(1-\dfrac{x}{2}\right)}$$

$$\equiv 2(1+x)^{-1} - \dfrac{3}{2}\left(1-\dfrac{x}{2}\right)^{-1}$$

Now, $(1+x)^{-1} = 1 - x + x^2 + \ldots$ and $\left(1-\dfrac{x}{2}\right)^{-1} = 1 + \dfrac{x}{2} + \dfrac{x^2}{4} + \ldots$

Overall, then, we conclude that:

$$\dfrac{5x-1}{(x+1)(x-2)} \approx 2(1-x+x^2) - \dfrac{3}{2}\left(1+\dfrac{x}{2}+\dfrac{x^2}{4}\right) = \dfrac{1}{2} - \dfrac{11}{4}x + \dfrac{13}{8}x^2 \text{ up to the term in } x^2.$$

When considering the values of x for which this series is valid, we select the more restricted range from the two series: the first gives $|x| < 1$, the second $|x| < 2$. For the overall series, $|x| < 1$.

19

MATHEMATICS: ANALYSIS AND APPROACHES HL

For the function $f(x) = \dfrac{2x-1}{(1-x)(3+x)}$, $x \ne 1$ and $x \ne -3$

(a) Show that $f(x)$ can be written as $\dfrac{1}{4(1-x)} - \dfrac{7}{4(3+x)}$.

(b) Obtain the first three terms of the expansion of $f(x)$, stating the values of x for which the expansion is valid.

(c) Find the percentage error when using the expansion to approximate to $f(x)$ when $x = 0.1$.

(a) $\dfrac{2x-1}{(1-x)(3+x)} \equiv \dfrac{A}{1-x} + \dfrac{B}{3+x}$

$2x - 1 \equiv A(3+x) + B(1-x)$

Put $x = -3$: $-7 = 4B \Rightarrow B = -\dfrac{7}{4}$

Put $x = 1$: $1 = 4A \Rightarrow A = \dfrac{1}{4}$

Thus $\dfrac{2x-1}{(1-x)(3+x)} \equiv \dfrac{1}{4(1-x)} - \dfrac{7}{4(3+x)}$

(b) $(1-x)^{-1} \approx 1 + x + x^2$

$\dfrac{1}{3}\left(1 + \dfrac{x}{3}\right)^{-1} \approx \dfrac{1}{3}\left(1 - \dfrac{x}{3} + \dfrac{x^2}{9}\right)$

$f(x) \approx \dfrac{1}{4}(1 + x + x^2) - \dfrac{7}{12}\left(1 - \dfrac{x}{3} + \dfrac{x^2}{9}\right)$

$= -\dfrac{1}{3} + \dfrac{4x}{9} + \dfrac{5x^2}{27}$ for $|x| < 1$

(c) $f(0.1) = -0.286738...$, Expansion $= -0.287037...$

% error $= \dfrac{|(-0.286738) - (-0.287037)|}{|-0.286738|} \times 100 = 0.104\%$

The principles here are exactly the same as in the example above, but the maths is a bit harder because of the more complicated fractions and the multitude of minus signs. It's the sort of question which is worth doing in rough first to avoid an answer which contains lots of crossings out. Just take each stage carefully, and check each bit of the answer before carrying on.

In part (c) I've used modulus signs, but you would get the same answer if you simply treated the two values as positive numbers when calculating the error.

1.10 Complex Numbers

The square root of −1: The term "imaginary number" is off-putting because it makes $\sqrt{-1}$ seem very abstract. Remember that negative numbers do not exist in real life either: you cannot have a negative amount of anything. As with negative numbers, once $\sqrt{-1}$ has been defined, we can fit it into our existing algebra - and very useful it turns out to be.

$x^2 + 2x + 5 = 0$
$x = \dfrac{-2 \pm \sqrt{4 - 20}}{2}$
$x = \dfrac{-2 \pm \sqrt{-16}}{2}$
$= \dfrac{-2 \pm 4i}{2} \Rightarrow x = -1 \pm 2i$

Definitions: The notes box shows the solution of $x^2 + 2x + 5 = 0$. The two solutions, $-1 + 2i$ and $-1 - 2i$ are called *complex numbers* and cannot be further simplified. A complex number z has the form $a + ib$, where a is called the *real part* and b is the *imaginary part*. These can be denoted by Re z and Im z. Note that all real numbers can be considered as complex numbers with a zero imaginary part. The *conjugate* of $z = a + ib$ is $z^* = a - ib$: thus complex roots of quadratic equations always occur in conjugate pairs.

Your GDC has an i button, so you can do all complex number calculations. Set in complex mode, it will also give complex solutions to equations.

Basic arithmetic: To **add** complex numbers, add their real parts together, then add their imaginary parts.

$(a + ib) + (c + id) = (a + c) + i(b + d)$

Example: $(2 + 3i) + (4 - i) = 6 + 2i$

Subtraction is similar:

$(a + ib) - (c + id) = (a - c) + i(b - d)$

Example: $(4 + 6i) - (5 + 3i) = -1 + 3i$

Multiplication uses the expansion of brackets:

$(a + ib) \times (c + id) = ac + iad + ibc + i^2bd = (ac - bd) + i(ad + bc)$

Never forget that $i^2 = -1$

Example: $(2 + 4i) \times (2 + 3i) = 4 + 6i + 8i + 12i^2 = -8 + 14i$

One useful fact that follows is that both the addition and multiplication of conjugate pairs give real results.

$(a + ib) + (a - ib) = 2a$ (or $z + z^* = 2\text{Re}(z)$)

$(a + ib) \times (a - ib) = a^2 - i^2b^2 = a^2 + b^2$ (or $zz^* = |z|^2$)

The second result leads to the method for division of complex numbers – similar to the rationalisation of surds.

To *divide* complex numbers:

1. Write the division as a fraction
2. Multiply top and bottom by the conjugate of the bottom
3. Simplify to the form $a + ib$

Beware:
The conjugate of $3i$ is $-3i$
The conjugate of 3 is 3
The conjugate of $2i + 1$ is $-2i + 1$
Always change the sign of the imaginary part

Example: Divide $3 + 2i$ by $1 - 4i$

$$\frac{3 + 2i}{1 - 4i} = \frac{(3 + 2i)(1 + 4i)}{(1 - 4i)(1 + 4i)} = \frac{-5 + 14i}{17} = \frac{-5}{17} + \frac{14}{17}i$$

Equality of complex numbers: If two complex numbers are equal then their real parts can be equated, and their imaginary parts can be equated, often resulting in two equations for the price of one. Use this to solve equations involving complex numbers.

Example: Solve $z^2 = 21 + 20i$, where Re z and Im z are both real numbers.

Solution: It often helps (but not always!) when solving questions involving complex algebra to write z as $a + bi$. So we get:

See website for the full working

$(a + bi)^2 = 21 + 20i \Rightarrow a^2 - b^2 + 2abi = 21 + 20i$

Now we can equate real and imaginary parts and, since this leads to two equations, we can solve simultaneously.

$a^2 - b^2 = 21$ and $2ab = 20 \Rightarrow b = \pm 2$ or $\pm 5i$

But b is real, so $b = \pm 2$ and $a = \pm 5$ giving $z = 5 + 2i$ or $-5 - 2i$

MATHEMATICS: ANALYSIS AND APPROACHES HL

> The complex number z satisfies $\dfrac{z}{z+2} = 2 - i$. Find:
>
> (a) z
>
> (b) $\arg(z)$ in an exact form
>
> (c) $|z + 2|$

(a) Let $z = x + iy$

$\dfrac{x + iy}{(x + 2) + iy} = 2 - i$

$x + iy = (2 - i)((x + 2) + iy)$

$x + iy = 2x + 4 + 2iy - ix - 2i - i^2 y$

$x + iy = (2x + y + 4) + i(2y - x - 2)$

Equating real parts and imaginary parts:

$x = 2x + y + 4 \Rightarrow x + y = -4$

$y = 2y - x - 2 \Rightarrow x - y = -2$

Solving simultaneously, $x = -3$ and $y = -1$

Thus $z = -3 - i$

(b) $\arg(z) = \pi + \arctan\left(\dfrac{1}{3}\right)$

(c) $|z + 2| = |-3 - i + 2|$

$= |-1 - i|$

$= \sqrt{2}$

Why did I use x and y instead of a and b? Why not?!

The most important point to note in the working is that wherever a complex number with several terms has appeared, I have collected together real and imaginary parts – see particularly lines 2 and 5 of part (a).

Parts (b) and (c) relate to modulus-argument form, covered in the next section.

Once again, in part (c), I have written z as x + iy and then, in line 2, collected together real and imaginary parts.

> The fact that a polynomial of degree n has n roots is known as the *Fundamental Theorem of Algebra*.

Roots of polynomial equations: How many roots are there to the polynomial equation $P(x) = 0$ if $x \in \mathbb{R}$? If $P(x)$ is a quadratic then there can be either 0 or 2 roots (a repeated root counts as 2). If it's a cubic, there can be 1 or 3, and so on. But if $x \in \mathbb{C}$ then a quadratic will always have 2 (both real or a conjugate pair); a cubic will always have 3 (all real or 1 real and a conjugate pair). In general, a degree n equation has n roots, and any complex roots will always occur in conjugate pairs. Thus, if $2 - 3i$ is a root of $P(x) = 0$, then $2 + 3i$ must be another root.

Example: Find a cubic equation (with real coefficients) which has roots $(2 - i)$ and 3.

Solution: The three roots will be 3, $2 - i$ and $2 + i$ so the equation can be written in factor form as $(x - 3)(x - (2 - i))(x - (2 + i)) = 0$

This multiplies out to give $(x - 3)(x^2 - 4x + 5) = 0$ and so the cubic equation is $x^3 - 7x^2 + 17x - 15 = 0$

> To multiply out the brackets, first remove the inner brackets to get:
> $(x - 2 + i)(x - 2 - i)$

> The quadratic factor can be found more easily using the sum of roots method – see page 57.

1.11 The Complex Plane

Argand diagrams: We can use a *number line* to interpret real numbers geometrically. Each number is represented by a point on the line, or by a position vector – see the number 2 below.

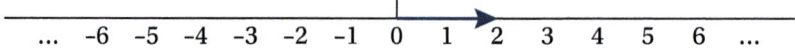

22

1. NUMBER AND ALGEBRA

The operation of multiplication by –1 can be represented geometrically by a rotation of 180°. Since $i^2 = -1$, multiplication by i can therefore be represented by a 90° rotation.

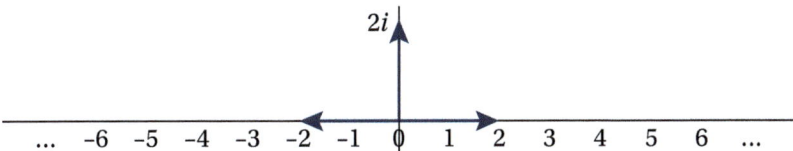

This operation puts $2i$ along a new number line at right angles to the real number line. This is called the *imaginary* axis, and explains why you cannot "see" complex numbers when just operating with real numbers. If we now use the two number lines (real and imaginary) to form two axes, the result is the *complex plane*. Complex numbers can be represented by points or vectors, thus forming *Argand diagrams*.

Magnitude of a complex number: The Argand diagram gives us a way of comparing the size of complex numbers to the size of real numbers. Clearly the numbers 2 and –2 have size (or modulus) 2; this is also the length of their vector representations. The vectors representing $2i$ and $3 + 2i$ have lengths 2 and $\sqrt{13}$: using modulus notation we can write $|3 + 2i| = \sqrt{13}$. In general $|a + bi| = \sqrt{a^2 + b^2}$.

Note that $zz^* = (a + bi)(a - bi)$ which expands to $a^2 + b^2$. Thus $zz^* = |z|^2$.

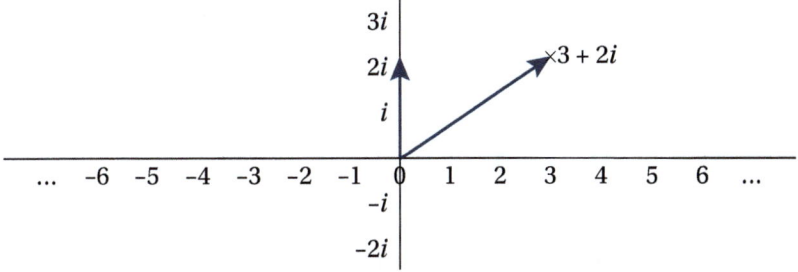

***r*cosθ + *ir*sinθ form**: Looking at the vector representation above suggests that a complex number could conveniently be written as a length (*modulus*) and angle (*argument*). The argument is the angle made with the real axis, and is usually written such that it lies between $-\pi$ and π (although it can also be expressed in degrees). Thus, $z = 3 + 2i$ has length $\sqrt{13}$ and angle $\arctan(2/3) \approx 33.7°$. Working backwards, this means that $\text{Re}(z) = \sqrt{13} \cos 33.7°$ and $\text{Im}(z) = \sqrt{13} \sin 33.7°$.

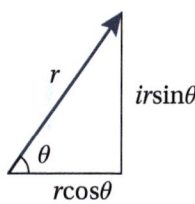

We now have an alternative form for z:

$$3 + 2i = \sqrt{13} \cos 33.7° + i\sqrt{13} \sin 33.7° = \sqrt{13}(\cos 33.7° + i \sin 33.7°)$$

This can be written in shorthand as $\sqrt{13} \text{ cis } 33.7°$. This form is particularly useful for multiplication and division since:

- $r \text{cis} \, \theta \times s \text{cis} \, \phi = rs \text{cis}(\theta + \phi)$ *(Multiply moduli, add arguments)*
- $r \text{cis} \, \theta \div s \text{cis} \, \phi = \frac{r}{s} \text{cis}(\theta - \phi)$ *(Divide moduli, subtract arguments)*

Note that conversion from $a + ib$ form to $r \cos \theta + ir \sin \theta$ form is the same as converting Cartesian to polar coordinates – make sure you can do this on your calculator (and the reverse as well).

In summary:

- $|a + ib| = \sqrt{(a^2 + b^2)}$
- $\arg(a + ib) = \tan^{-1}(b/a)$ *if in quadrants 1 or 2, otherwise add π*
- $\text{Re}(r \text{cis} \theta) = r \cos \theta$
- $\text{Im}(r \text{cis} \theta) = r \sin \theta$

> Find out how to set your GDC to calculate either in $a + bi$ mode or in $re^{i\theta}$ mode. As ever, also check that you are in degrees or radians as appropriate.

An alternative form for $r\text{cis}\theta$ is $re^{i\theta}$. You don't need to know the derivation of this, but must recognise that, for example, the complex number $3e^{\frac{\pi}{3}i}$ has modulus 3 and argument $\frac{\pi}{3}$.

Example: Write $\frac{10 - 5i}{3 - 4i}$ in modulus-argument form.

Solution: $10 - 5i = \sqrt{125}\,\text{cis}(-0.4636)$ and $3 - 4i = 5\text{cis}(-0.9273)$

Thus $\frac{10 - 5i}{3 - 4i} = \frac{\sqrt{125}}{5}\,\text{cis}(-0.4636 - (-0.9273))$

$= \sqrt{5}\,\text{cis}(0.464)$

> An alternative method is to divide the numbers in $a + bi$ form and then convert to modulus-argument form. Try this for yourself and decide which method is quicker.

Complex numbers as vectors: An Argand diagram represents a complex number as a vector. For example, $3 - 4i$ is equivalent to the vector $\begin{pmatrix} 3 \\ -4 \end{pmatrix}$. Addition of complex numbers can be represented by the addition of vectors:

$(3 - 4i) + (-1 + 6i) = (2 + 2i)$ is equivalent to $\begin{pmatrix} 3 \\ -4 \end{pmatrix} + \begin{pmatrix} -1 \\ 6 \end{pmatrix} = \begin{pmatrix} 2 \\ 2 \end{pmatrix}$.

Multiplication of complex numbers also has a geometric transformational equivalence, but rather less obviously. Multiplying z by the complex number $r\text{cis}\theta$ is equivalent to multiplying $|z|$ by r, and then rotating by θ. Division, of course, will divide $|z|$ by r and rotate by $-\theta$.

1.12 De Moivre's Theorem

Definition: De Moivre's Theorem gives us a quick way of raising a complex number to any power. To use it, the complex number must be written in $r\text{cis}\theta$ form. Then:

$(r\text{cis}\theta)^n = r^n\text{cis}n\theta$

> We can also demonstrate De Moivre's theorem using the exponential form:
> $(re^{i\theta})^n = r^n e^{in\theta}$

When written in full, this appears as:

$(r\cos\theta + ir\sin\theta)^n = r^n\cos n\theta + ir^n\sin n\theta$

It is often used with $r = 1$, in which case:

$(1\text{cis}\theta)^n = 1\text{cis}n\theta$

or $(\cos\theta + i\sin\theta)^n = \cos n\theta + i\sin n\theta$

Applications: One simple use is to find high powers of complex numbers written in $a + ib$ form.

> Although questions usually expect answers in radians, you can use degrees to make the working easier if you want.
>
> See page 76 for the sin and cos of common angles.

Example: Calculate $(1 + i\sqrt{3})^4$ giving the answer in $a + bi$ form.

Solution: Rewrite in mod-arg form: $\left(2\text{cis}\left(\frac{\pi}{3}\right)\right)^4$

Use De Moivre's theorem: $16\,\text{cis}\left(\frac{4\pi}{3}\right)$

Convert back to $a + bi$ form: $16\cos\left(\frac{4\pi}{3}\right) + 16i\sin\left(\frac{4\pi}{3}\right)$

$= -8 - 8\sqrt{3}\,i$

You will find the majority of questions involve finding the roots of complex numbers. First, let's see how we can use De Moivre's theorem to find the roots of 1.

1. NUMBER AND ALGEBRA

Roots of unity: 1 has two square roots, 1 and −1. In the real numbers, it has only one cube root, but we can also find two complex roots. First we write 1 in polar form as 1cis0, but note that we can also repeatedly add 2π giving: $1 = 1\operatorname{cis}0$ or $1\operatorname{cis}2\pi$ or $1\operatorname{cis}4\pi$.

Using complex numbers, every number will have 2 square roots, 3 cube roots and so on.

Now let's form the equation $z^3 = 1 \Rightarrow z = 1^{\frac{1}{3}}$. Thus:

$$z = 1^{\frac{1}{3}} = (1\operatorname{cis}0)^{\frac{1}{3}} \text{ or } (1\operatorname{cis}2\pi)^{\frac{1}{3}} \text{ or } (1\operatorname{cis}4\pi)^{\frac{1}{3}}$$

Using De Moivre's theorem (which applies to all powers), we get:

$$z = 1^{\frac{1}{3}} = 1\operatorname{cis}0 \text{ or } 1\operatorname{cis}\left(\frac{2\pi}{3}\right) \text{ or } 1\operatorname{cis}\left(\frac{4\pi}{3}\right).$$

Strictly, $1\operatorname{cis}\left(\frac{4\pi}{3}\right)$ should be written as $1\operatorname{cis}\left(-\frac{2\pi}{3}\right)$.

Converting these into $a + ib$ form we find the cube roots of 1 are:

$$1, -0.5 + \frac{i\sqrt{3}}{2}, -0.5 - \frac{i\sqrt{3}}{2}$$

Note the following points:

- Each root has magnitude 1, so on an Argand diagram each root is on a circle, radius 1, centre the origin.
- If we were to work out the next root in the series we would get $1\operatorname{cis}\left(\frac{6\pi}{3}\right)$ which is 1 again – so there are only 3 roots.
- Symmetry can be used to find all the roots, given the first.

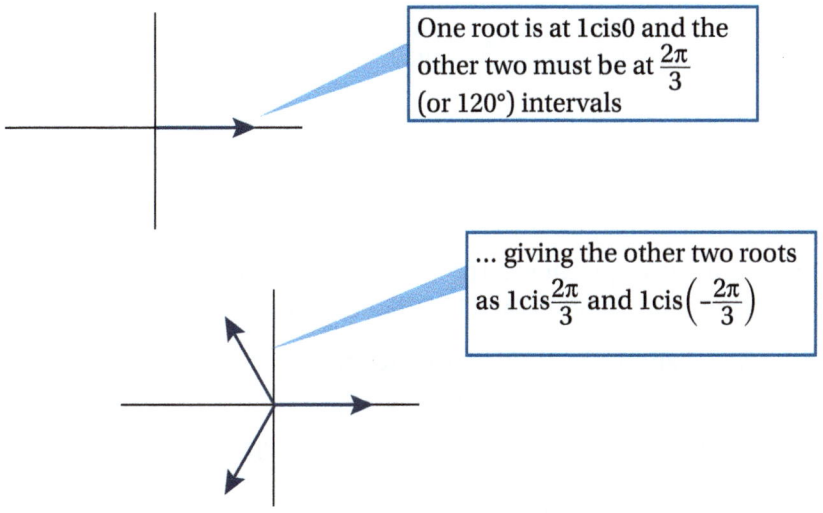

One root is at 1cis0 and the other two must be at $\frac{2\pi}{3}$ (or 120°) intervals

... giving the other two roots as $1\operatorname{cis}\frac{2\pi}{3}$ and $1\operatorname{cis}\left(-\frac{2\pi}{3}\right)$

Cube roots of complex numbers will always be spaced at 120°, fourth roots at 90°, fifth roots at 72° and so on.

Roots of complex numbers: The method can now be extended to find the roots of any real or complex number.

Example: Find all the solutions of $z^5 = 1 + i$.

Solution: $1 + i = \sqrt{2}\operatorname{cis}45°$, so we must solve $z = (\sqrt{2}\operatorname{cis}45)^{\frac{1}{5}}$

De Moivre's theorem gives one solution, $z_1 = \sqrt[10]{2}\operatorname{cis}9°$. The other solutions will then be spaced at 72° intervals, giving:

$z_2 = \sqrt[10]{2}\operatorname{cis}81°, \sqrt[10]{2}\operatorname{cis}153°, \sqrt[10]{2}\operatorname{cis}(-135°), \sqrt[10]{2}\operatorname{cis}(-63°)$. In $a + bi$ form:

$1.06 + 0.17i, 0.17 + 1.06i, -0.95 + 0.49i, -0.76 - 0.76i, 0.49 - 0.95i$.

I've used degrees for convenience of calculation. Make sure you can convert between both forms of complex numbers using your GDC.

MATHEMATICS: ANALYSIS AND APPROACHES HL

We now have a library of techniques for solving complex number questions and, occasionally, can be spoilt for choice. Look at the next question, for example:

Express $\dfrac{1}{(1-i\sqrt{3})^3}$ in the form $\dfrac{p}{q}$, where $p, q \in \mathbb{Z}$.	
The answer is $-\dfrac{1}{8}$. I suggest four possible methods below – it would be a good idea for you to work through each of them to get a feel for the possibilities when faced with an exam question.	As a general rule, multiplications and divisions are easier using $r\operatorname{cis}\theta$ form; but additions and subtractions have to be done in $a + bi$ form.
Method 1: Multiply out the denominator as three brackets using ordinary algebra.	
Method 2: Use binomial expansion to multiply out the denominator.	
Method 3: Write 1 and $1 - i\sqrt{3}$ in $r\operatorname{cis}\theta$ form. Use De Moivre's theorem to simplify the denominator.	
Method 4: Write $1 - i\sqrt{3}$ in $r\operatorname{cis}\theta$ form then use De Moivre's theorem to simplify $(1 - i\sqrt{3})^{-3}$.	

You can find full working for these four methods on the website.

Complex Numbers: Practice Exercise

Answers

1. $6 - i$, $4 + 5i$, $11 - 13i$, $-\dfrac{1}{10} + \dfrac{17}{10}i$
2. $5 + 5i$
3. $2\operatorname{cis}\left(-\dfrac{2\pi}{3}\right)$
4. $2 + 2i$
5. $\pm(2 + i)$
6. $-1, -1 - i\sqrt{2}$
7. $-2, 1 \pm i\sqrt{3}$
8. $5\theta = 5\cos^4\theta\sin\theta - 10\cos^2\theta\sin^3\theta + \sin^5\theta$

These aren't exam-style questions. The aim of this exercise is for you to practise the various complex number methods and techniques which are used in longer questions.

1. For $z_1 = 5 + 2i$ and $z_2 = 1 - 3i$, calculate: $z_1 + z_2$, $z_1 - z_2$, $z_1 z_2$, $\dfrac{z_1}{z_2}$. Also show that $|z_1 z_2| = |z_1||z_2|$ and $\arg(z_1 z_2) = \arg z_1 + \arg z_2$.

2. Given that $\dfrac{1}{1+2i} + \dfrac{1}{2+i} = \dfrac{6}{z}$, find z in the form $a + bi$.

3. Convert $-1 - i\sqrt{3}$ to $r\operatorname{cis}\theta$ form.

4. Convert $\sqrt{8}\, e^{\frac{i\pi}{4}}$ to $a + bi$ form. (Both with and without GDC.)

5. Use the equality of complex numbers to find the square roots of $3 + 4i$.

6. Given that $-1 + i\sqrt{2}$ is a root of $x^3 + 3x^2 + 5x + 3 = 0$, find the other two roots.

7. Draw the three cube roots of -8 on an Argand diagram. Write each of them in $a + bi$ form, and show that they add to 0.

8. Use the binomial theorem to expand $(\cos\theta + i\sin\theta)^5$. Hence use De Moivre's theorem, and the equality of complex numbers, to find an expression for $\sin 5\theta$ in terms of $\sin\theta$ and $\cos\theta$.

1. NUMBER AND ALGEBRA

1.13 Deductive proof

Equality and identity: We often use an "equals" sign rather sloppily. Consider the following two statements:

$$3(x - 2) = 2x \qquad (1)$$
$$3(x - 2) = 3x - 6 \qquad (2)$$

Statement 1 is an *equation* and we are looking for solutions. In this case the only solution is $x = 6$. In other words, LHS and RHS will only be equal when $x = 6$. Statement 2 is an *identity*. The two expressions always have the same value whatever the value of x: they are just different ways of writing the same thing. Identities should always be written using the appropriate symbol which is $\equiv$.

LHS and RHS stand for left-hand side and right-hand side

$$3(x - 2) \equiv 3x - 6$$

LHS to RHS proof: Deductive proof is often called LHS to RHS proof to underline the fact that we must work *from* the LHS, ending up with the RHS. Solving an equation can be carried out by performing the same operations on both sides of the equation. You can't do this to prove an identity because you would need to start with the thing you are trying to prove! To take a simple example, let's prove $-1 = 1$.

Suppose $\qquad -1 = 1$

Square both sides: $(-1)^2 = 1^2$

$\qquad\qquad\qquad 1 = 1$

Since LHS = RHS, the original statement must be true!

This doesn't work because the algebra isn't necessarily reversible. Because $x^2 = y^2$ it doesn't follow that $x = y$.

So, never start a deductive proof by writing down what you want to prove – always start with *just* the LHS.

Example: Show that $(x - 2)^2 - 5 \equiv x^2 - 4x - 1$

Solution: $(x - 2)^2 - 5 \equiv x^2 - 4x + 4 - 5$

$\qquad\qquad\qquad \equiv x^2 - 4x - 1 \qquad$ qed

One equals sign per line, and align them as you go down.

qed: quod erat demonstrandum (which was to be shown).

When asked to prove conjectures involving numbers, remember that an even number is always of the form $2k$, and an odd number of the form $2k + 1$. For example, here is a proof that the sum of the squares of two consecutive integers is always an odd number:

Let the two integers be n and $n + 1$.

The sum of the squares $= n^2 + (n + 1)^2$

$\qquad\qquad\qquad\qquad = n^2 + n^2 + 2n + 1$

$\qquad\qquad\qquad\qquad = 2n^2 + 2n + 1$

$\qquad\qquad\qquad\qquad = 2(n^2 + n) + 1$

This is of the form $2k + 1$ and so is always odd.

MATHEMATICS: ANALYSIS AND APPROACHES HL

> (a) Show that $\dfrac{1}{m-1} + \dfrac{1}{m^2-m} \equiv \dfrac{m+1}{m^2-m}$
>
> (b) Show that the identity is valid when $m = 2$

(a) $\dfrac{1}{m-1} + \dfrac{1}{m^2-m} \equiv \dfrac{1}{m-1} + \dfrac{1}{m(m-1)}$

$\equiv \dfrac{m}{m(m-1)} + \dfrac{1}{m(m-1)}$

$\equiv \dfrac{m+1}{m(m-1)}$

$\equiv \dfrac{m+1}{m^2-m}$ qed

(b) When $m = 2$,

LHS $= \dfrac{1}{1} + \dfrac{1}{2} = \dfrac{3}{2}$

RHS $= \dfrac{3}{4-2} = \dfrac{3}{2}$

∴ LHS = RHS

Make sure you know how to deal with algebraic fractions. As with numerical fractions, it is likely that you will need to find a common denominator.

> A triangle has sides $\sqrt{2n+1}$, n and $n+1$, where $n+1$ is the longest side. Prove that the triangle is right-angled for any value of n.

We need to show that $(\sqrt{2n+1})^2 + n^2 \equiv (n+1)^2$

$(\sqrt{2n+1})^2 + n^2 \equiv 2n + 1 + n^2$

$\equiv n^2 + 2n + 1$

$\equiv (n+1)(n+1)$

$\equiv (n+1)^2$

∴ The triangle is right-angled

Right-angled triangle? Let's use Pythagoras' Theorem.

There is a temptation to multiply out the RHS, but in LHS to RHS proofs you should work on the LHS only – unless you really can't see any way to get to the RHS unless you perform algebra on it as well.

Note that I finished with the statement we were asked to prove

Deductive proof: Practice Exercise

Here's a variety of proof questions. I suggest that in each case you also try substituting a value and show that LHS = RHS for that value.

1. Prove that $(n+1)^2 - (n-1)^2 \equiv 4n$

2. Prove that $\dfrac{x^2+x}{x^2+4x+3} \equiv \dfrac{x}{x+3}$

3. Given that k is any integer write down, in terms of k, three consecutive integers. Hence show that the sum of three consecutive integers is always a multiple of 3.

4. The nth term in the sequence of triangle numbers 1, 3, 6, 10, ….is given by $\tfrac{1}{2}n(n+1)$. Prove that 8 times any triangle number is always 1 less than a square number.

1. NUMBER AND ALGEBRA

1.14 Proof by Induction

Induction is used to prove a general result on the set of natural numbers which has already been obtained by surmise or guesswork. A question could begin: "Prove by induction that, for $n > 0$, ..." The method to follow is:

- show the result is true for the first value of n (usually 1);
- then show that if the result holds for some general value (say, k) then it will also hold for $k + 1$. This will require some means of moving from the kth result to the $(k + 1)$th, depending on the type of question.

> The second part often involves heavy algebra: don't then forget the first part which is usually a simple substitution.

The most common proofs you are asked to carry out are:

- General formula for the sum of a series
- Divisibility testing
- General algebraic formulae (such as de Moivre's Theorem) and also in calculus and trigonometry.

These are best illustrated by some worked examples. The wording I have used in the first two examples illustrates what an examiner might expect. I have been more informal in later examples.

Example: Prove that $1 \times 3 + 2 \times 4 + ... + n(n + 2) = \frac{1}{6}n(n + 1)(2n + 7)$ for $n \geq 1$

Solution: Let P(n) be the proposition $1 \times 3 + 2 \times 4 + ... + n(n + 2) = \frac{1}{6}n(n + 1)(2n + 7)$

First we show that the formula works for n = 1.

When $n = 1$, $1 \times 3 = 3$. $\frac{1}{6}n(n + 1)(2n + 7) = \frac{1}{6} \times 1 \times 2 \times 9 = 3$

∴ P(1) is true

We now need to show that if the formula is true for some value k, it is also true for k + 1. The trick is to see what we are aiming for – in this case, the formula for n = k + 1. This is $\frac{1}{6}(k + 1)(k + 2)(2k + 9)$. So, Let's add on the next term in the series and see what happens.

Assume P(k) is true: $1 \times 3 + 2 \times 4 + ... + k(k + 2) = \frac{1}{6}k(k + 1)(2k + 7)$

Then, for $n = k + 1$:

$1 \times 3 + 2 \times 4 + ... + k(k + 2) + (k + 1)(k + 3) = \frac{1}{6}k(k + 1)(2k + 7) + (k + 1)(k + 3)$

Now all we have to do is show that the RHS is the same as what we are aiming for. You could multiply out both expressions, but it is better to use factorisation:

$\frac{1}{6}k(k + 1)(2k + 7) + (k + 1)(k + 3) = \frac{1}{6}(k + 1)[k(2k + 7) + 6(k + 3)]$

$= \frac{1}{6}(k + 1)(2k^2 + 13k + 18)$

$= \frac{1}{6}(k + 1)(k + 2)(2k + 9)$

Note that we had to insert a 6 in front of the second term in the brackets to allow us to take the fraction outside the brackets. Also, a sneak look at our aiming point makes the quadratic factorisation easy.

P(k) is true ⇒ P(k + 1) is true, and P(1) is true.

Hence, by induction, P(n) is true for $n \geq 1$.

Try exactly the same technique to prove:

(a) $1^2 + 2^2 + 3^2 + 4^2 + \ldots + n^2 = \frac{1}{6}n(n+1)$

(b) $\frac{1}{1 \times 2} + \frac{1}{2 \times 3} + \frac{1}{3 \times 4} + \ldots + \frac{1}{n(n+1)} = \frac{n}{n+1}$

(c) $\sum_{k=1}^{n} 2^{k-1} = 2^n - 1$

You can find a video solution for a question similar to part (c) on the website

But proving a formula for the sum of a series is just one of the tricks you must perform. Another type of question is a divisibility test. Here, the aiming point is to show that if u_k is divisible by the number, use this to show that u_{k+1} is as well.

Example: Prove that $4^n + 6n - 1$ is divisible by 9 for $n \geq 1$.

Solution: Let P(n) be the proposition that $4^n + 6n - 1$ is divisible by 9 for $n \geq 1$.

When $n = 1$, $4^1 + 6 - 1 = 9$, which is divisible by 9. So P(1) is true.

The ($k + 1$)th term will be $4^{k+1} + 6k + 5$, and we need to get there from the kth term. Here goes:

Assume P(k) is true. A number divisible by 9 is of the form 9t.

$4^k + 6k - 1 = 9t$ Now multiply both sides by 4 to get $4k + 1$

$4^{k+1} + 24k - 4 = 36t$ Now subtract 18k to get 6k

$4^{k+1} + 6k - 4 = 36t - 18k$ Now add 9 to get 5

$4^{k+1} + 6k + 5 = 36t - 18k + 9$

$4^{k+1} + 6k + 5 = 9(4t - 2k + 1)$ which is divisible by 9

So P(k) is true $\Rightarrow$ P($k + 1$) is true, and P(1) is true.

Hence, by induction, P(n) is true.

Now try proving by induction that:

(a) 6^{n-1} is divisible by 5 for $n \geq 1$

(b) $8^n - 7n + 6$ is divisible by 7 for $n \geq 1$

Example: Prove that $\frac{d}{dx}(x^n) = nx^{n-1}$

Solution: I'll leave it to you to show that P(1) is true. Now, what is our "aiming point"? It must be to show that $\frac{d}{dx}(x^{k+1}) = (k+1)x^k$. And how can we use P($k$) to get to P($k + 1$)? The product rule provides the answer:

If P(k) is true then $\frac{d}{dx}(x^k \times x) = x^k \times 1 + kx^{k-1} \times x$

$= x^k + kx^k$

$= (k + 1)x^k$

As an exercise, write out the whole proof formally.

You can find the full working on the website.

Now try and prove De Moivre's theorem using induction.

Occasionally an inductive proof involves an inequality. For example, prove that $5^k + 9 < 6^k$ for $n \geq 2$. Our aim is to show that $5^{k+1} + 9 < 6^{k+1}$ and this looks like the divisibility proofs. The difference is that we can't make the expressions exactly the same – but we don't need to.

1. NUMBER AND ALGEBRA

Here's the key part of the working:

$$5^k + 9 < 6^k$$
$$6(5^k + 9) < 6^{k+1}$$
$$(5+1)(5^k + 9) < 6^{k+1}$$
$$5^{k+1} + 5^k + 45 + 9 < 6^{k+1}$$
$$(5^{k+1} + 9) + 5^k + 45 < 6^{k+1}$$

Now, since $5^k + 45 > 0$ we can subtract it from the LHS and the inequality must still be true. Hence $P(k) \Rightarrow P(k+1)$.

1.15 Proof by Contradiction

We have seen proof by deduction (starts with a premise) and proof by induction (starts with the conclusion). Proof by contradiction starts by assuming the opposite proposition is true, and then showing that this leads to a contradiction. A classic proof is to show that $\sqrt{2}$ is irrational. We begin by assuming that it is rational (in other words, it can be written as a fraction in its simplest form).

So, let us assume that $\sqrt{2} = \frac{p}{q} \Rightarrow 2q^2 = p^2$. This means that p^2 is even and hence p is even. Thus p can be written as $2k$. Substituting, $2q^2 = (2k)^2$ giving $2q^2 = 4k^2 \Rightarrow q^2 = 2k^2$. Thus q^2 is even, and q is therefore even. But if both p and q are even, then $\frac{p}{q}$ is not in its simplest form. The proposition leads to a contradiction and hence has been proved to be false. Therefore, $\sqrt{2}$ is irrational.

1.16 Systems of Equations

A set of equations which is to be solved simultaneously is known as a *system of equations*. In general, 2 equations are required to find two unknowns, 3 equations to find 3 unknowns and so on. You should be familiar with the methods of *elimination* and *substitution* required to solve a pair of simultaneous equations. Similar methods can be used to solve a system of three equations, but the working can be quite long; an alternative method is known as *row reduction*. First, let's see how the familiar algebraic methods work.

Example: Solve the system of equations for x, y and z:

$$\begin{cases} 4x + y + 2z = 6 & \dots(1) \\ x - 3y - 4z = -1 & \dots(2) \\ 3x + 2y + 3z = 5 & \dots(3) \end{cases}$$

Solution: First we must eliminate one letter between two pairs of equations. Examining the numbers, I choose to eliminate y from equations 1 and 2, and then from equations 1 and 3. This gives:

$$\begin{cases} 13x + 2z = 17 \\ 5x + z = 7 \end{cases}$$

We are now in familiar territory, and can use either elimination or substitution to find that $x = 1$. Re-substituting gives $z = 2$, and then $y = -2$.

I did, however, choose the numbers to make it work quite easily – it can get very messy! You also need to try and work out in advance which are the best letters to eliminate. Row reduction gives us a method where you perform the same actions every time. Let's see how it works with the following system of equations:

$$\begin{cases} 2x + 4y - z = 12 \\ x - y + 4z = 6 \\ 4x + 5y - z = 17 \end{cases}$$

We're just going to work on the numbers, so begin by rewriting the equations without the variables – it helps to leave a gap before the final column. The aim now is to use *row operations* (that is, multiplications and subtractions of whole rows at a time) to reduce the three numbers in the bottom left corner to 0. The method is as follows:

1. First, use row operations to replace the left hand numbers in rows 2 and 3 (starred) with zeroes. You usually use the top left hand number to do this.

2. Next, use row operations to replace the bottom middle number (starred) with a zero. You **must** use the middle number to do this.

3. The bottom left hand triangle of numbers will now be zeroes. Replace x, y and z to form three equations which can be easily solved. Write the bottom equation first.

2	4	-1	12	
1*	-1	4	6	Setup
4*	5	-1	17	

2	4	-1	12	Leave R_1 unchanged
0	-6	9	0	R_2 becomes $2 \times R_2 - R_1$
0	-3*	1	-7	R_3 becomes $R_3 - 2 \times R_1$

2	4	-1	12	Leave R_1 unchanged
0	-6	9	0	Leave R_2 unchanged
0	0	-7	-14	R_3 becomes $2 \times R_3 - R_2$

$$\begin{cases} -7z = -14 \\ -6y + 9z = 0 \\ 2x + 4y - z = 12 \end{cases}$$

to get $z = 2$, $y = 3$ and $x = 1$.

Equations with non-unique solutions: In the example above there was a single, or unique, solution for x, y and z. It is also possible that there are no solutions, or an infinity of solutions. We shall see a geometrical interpretation of these cases in the chapter on vectors. What will happen in the row reduction method to indicate either of these possibilities? Quite simply, when we create the two zeroes in the bottom row, we will find that the third number also reduces to a zero. If the fourth number is non-zero, there will be no solutions; and if it is also zero, there will be infinite solutions.

...because $0x + 0y + 0z = 0$ has ∞ solutions, whereas $0x + 0y + 0z = k$ has none.

For example, if you use row operations to solve the following system of equations:

$$\begin{cases} -2x + y - 5z = 4 \\ 6x - 2y + 4z = -2 \\ -4x + y + z = -2 \end{cases}$$

... you end up with the following equations:

$$\begin{cases} -2x + y - 5z = 4 \\ y - 11z = 10 \\ 0x + 0y + 0z = 0 \end{cases}$$

There are therefore an infinite number of solutions.

1. NUMBER AND ALGEBRA

We can also give a general form for the solution. Suppose we let $z = t$, then:

$y = 10 + 11t$

$-2x + (10 + 11t) - 5t = 4 \Rightarrow x = 3t + 3$

So the general solution is: $x = 3t + 3, y = 11t + 10, z = t$.

Consider the system of equations $\begin{cases} x + y - 4z = 1 \\ 5x + 2y - 9z = 7 \\ 4x - 2y + az = b \end{cases}$

(a) Find the unique solution when $a = 8$ and $b = 10$.

(b) Find the value of a and the range of values of b for which there are no solutions.

(c) (i) Find the value of a and the value of b for which there are an infinite number of solutions.

  (ii) For these values, find the general solution.

1	1	-4	1
5	2	-9	7
4	-2	a	b

$R_2 = R_2 - 5R_1$

$R_3 = R_3 - 4R_1$

1	1	-4	1
0	-3	11	2
0	-6	$a + 16$	$b - 4$

$R_3 = R_3 - 2R_2$

1	1	-4	1
0	-3	11	2
0	0	$a - 6$	$b - 8$

Since the row reduction method is the same whether you are looking for 0, 1 or an infinite number of solutions, it is best to do the initial working with a and b in place. Once this has been done, just remember that for 0 solutions the bottom row must be 0 0 0 not 0, and for an infinite number of solutions the bottom row must be 0 0 0 0.

(a) For $a = 8$ and $b = 10$, the system of equations is:

$2z = 2$

$-3y + 11z = 2$

$x + y - 4z = 1$

Thus: $z = 1, y = 3, x = 2$

(b) $a = 6, b \neq 8$

(c) (i) $a = 6, b = 8$

  (ii) $z = t$

  $-3y + 11z = 2 \Rightarrow y = \dfrac{11t - 2}{3}$

  $x + \dfrac{11t - 2}{3} - 4t = 1$

  $3x + 11t - 2 - 12t = 3$

  $x = \dfrac{5 + t}{3}$

To find the general solution, set $z = t$. Then use the reduced equations to find y and x.

The general solution is $x = \dfrac{5 + t}{3}, y = \dfrac{11t - 2}{3}, z = t$

MATHEMATICS: ANALYSIS AND APPROACHES HL

Number and Algebra: Long Answer Questions

On the following pages are a selection of section B style exam questions related to the Number and Algebra topic, although some of them may require knowledge and techniques from other areas of the syllabus. The answers are given here, but full working may be found on the Peak Study Resources website.

See www.peakib.com

1. NUMBER AND ALGEBRA

1. (a) (i) Find the number of multiples of 6 between 100 and 500.

   (ii) Hence find the probability that a number chosen at random between 100 and 500 inclusive will be a multiple of 6.

   (iii) Find the sum of the multiples of 6 between 100 and 500.

   (iv) Find the sum of the numbers between 100 and 500 which are multiples of 6 but *not* multiples of 8.

   (b) Prove by induction that $12^n + 2(5^{n-1})$ is a multiple of 7 for $n \in \mathbb{Z}^+$.

Answers:

(a) (i) 67   (ii) $\frac{67}{401}$ (iii) 20100   (iv) $20100 - 4800 = 15300$

2. $z = \frac{1}{2}e^{3i\theta}$ where $0 < \theta < \frac{1}{6}\pi$.

   (a) Find the modulus and argument of:

   (i) z

   (ii) z^*

   (iii) iz

   (b) Show that $(1+z)(1+z^*) = \cos 3\theta + \frac{5}{4}$

   (c) (i) Find the fifth roots of $4 + 4i$ in $re^{i\theta}$ form.

   (ii) Sketch all the roots on an Argand diagram.

   (iii) Find p and q such that $(p + qi)^5 = 4 + 4i$, where $p, q \in \mathbb{Z}$.

Answers:

(a) (i) $\frac{1}{2}, 3\theta$ (ii) $\frac{1}{2}, -3\theta$ (iii) $\frac{1}{2}, 3\theta + \frac{\pi}{2}$

(c) (i) $\sqrt{2}\,e^{\frac{i\pi}{20}}$, $\sqrt{2}\,e^{\frac{9i\pi}{20}}$, $\sqrt{2}\,e^{\frac{17i\pi}{20}}$, $\sqrt{2}\,e^{\frac{25i\pi}{20}}$, $\sqrt{2}\,e^{\frac{33i\pi}{20}}$

(ii)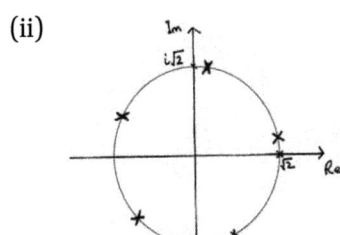

(iii) $p = -1, q = -1$

MATHEMATICS: ANALYSIS AND APPROACHES HL

3. $f(x) = \dfrac{x+5}{(x+2)(1-x)}$

   (a) Express $f(x)$ in the form $\dfrac{A}{2+x} + \dfrac{B}{1-x}$ where A and B are integers to be found.

   (b) Hence show that $f(x)$ can be approximated by the quadratic expression $g(x) = \dfrac{5}{2} + \dfrac{7}{4}x + \dfrac{17}{8}x^2$, and state the values of x for which the approximation is valid.

   (c) (i) Use your answer to (a) to find $\int f(x)\,dx$ giving your answer in the form of a single logarithm.

   (ii) Hence evaluate $\int_0^{0.5} f(x)\,dx$ giving your answer in an **exact** form.

   (d) (i) Evaluate $\int_0^{0.5} g(x)\,dx$ giving your answer to 4SF.

   (ii) Hence find the % error in using the approximation to find the value of the integral.

   ———————————

   Answers:

   (a) $\dfrac{1}{2+x} + \dfrac{2}{1-x}$

   (b) $|x| < 1$

   (c) (i) $\ln\dfrac{2+x}{(1-x)^2} + c$ or $\ln A\dfrac{2+x}{(1-x)^2}$ where $A = \ln c$ (ii) $\ln 5$

   (d) (i) 1.557 (ii) 3.24%

4. $1, a$ and b form the first three terms of a geometric progression G with common ratio k.

   $\log_2 x$, $\log_2 ax$ and $\log_2 bx$ form the first three terms of an arithmetic progression A.

   (a) For A,

   (i) Show that $2u_2 = u_1 + u_3$.

   (ii) Find the common difference in terms of k.

   (b) Show that, for A, $S_6 = 3\log_2(x^2 k^5)$.

   (c) Given that $S_6 = 33$ and $u_1 = 3$, find k.

   (d) (i) Show that $3, \dfrac{3}{a}, \dfrac{3}{b}$ are the first three terms of a geometric progression.

   (ii) Find the sum to infinity of the progression in (d)(i).

   ———————————

   Answers:

   (a) (ii) $\log_2 k$

   (c) $k = 2$

   (d) (ii) 6

Chapter 2: FUNCTIONS

2.1 Basics of Functions

A *relation* is an algebraic rule which shows how one set of numbers is related to, or obtained from another set. Relations often model real-life situations, so it is necessary to understand the different types of relation which occur and the notation used.

One-to-one and many-to-one: When two sets of numbers are related, their relationship can be defined in one of four ways:

- *One-to-one*: Each object has one image and vice versa.

    Examples:
    $x \to x + 1$
    Name of person → passport number

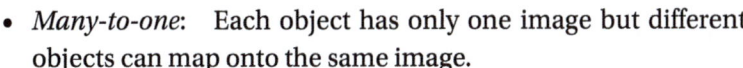

- *Many-to-one*: Each object has only one image but different objects can map onto the same image.

    Examples:
    $x \to$ nearest integer to x
    Student → mark in mathematics exam

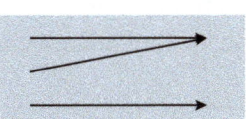

- *One-to-many*: An object can have more than one image but each image is related to only one object.

    Examples:
    $x \to \pm\sqrt{x}$
    Father → children

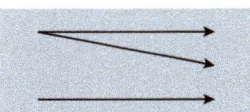

- *Many-to-many*: Each object can be related to several images and each image can be related to several objects.

    Examples:
    $x \to$ prime factors of x
    Vegetable → possible colours of vegetable

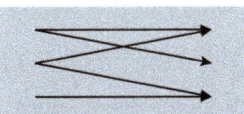

Functions are defined as relationships which are either one-to-one or many-to-one.

Function notation: A function is defined using the notation $f(x)$, but note that other letters may be used, particularly if modelling physical quantities. For example, the velocity of an object at a particular time may be defined using $v(t)$; the cost of buying a number of articles could be given by $C(n)$.

Thus the function $f(x) = 3x^2 - 1$ can be read as: "Function f takes any number, x, and turns it into $3x^2 - 1$." The function notation is also used with specific numbers; for example, $f(2) = 3 \times 2^2 - 1 = 11$.

Domain: The set of values to be input to a function is called the *domain*. In many functions, **any** value can be input, in which case the domain is $x \in \mathbb{R}$. However, the domain may be restricted for two reasons:

- Certain values of x may give impossible results, such as division by 0 or the square root of a negative number. For example, x cannot take the value 4 in the function $f(x) = \frac{x}{x-4}$, and this would be written as: $f(x) = \frac{x}{x-4}$ for $x \neq 4$.
- For the purposes of a particular question the domain may be "artificially" restricted. If $g(x) = 2x^2 - 3$ for $x > 0$, the function would only take positive values of x.

It's important to note the domain in an exam question because it may affect your answer. Solving $g(x) = 5$ in the example above would lead to the solution $x^2 = 4$ and hence $x = 2$; $x = -2$ is not in the domain.

Range: The set of values produced by a function is called the range. In the examples above, the range of g is $g(x) > -3$ since there is a minimum at $(0, -3)$. The range of f is best found by inspecting a graph: the range is the complete set of possible y values, so in this case would be $f(x) \neq 1$.

> If you draw the graph of f you will see there is a horizontal asymptote at $y = 1$. All other y values are represented on the graph.

The next sections use the image of a "function machine" which represents a function as a black box.

When the handle is turned, the 5 drops in the top and the function machine turns it into an 11!

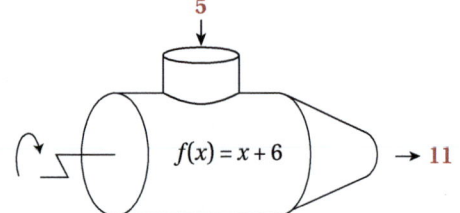

Inverse functions: An inverse function "reverses" the effect of a function. The inverse of add 2 is subtract 2. The inverse of squaring is square rooting. In terms of the function machine, just turn the handle the other way and the 11 turns back into a 5. The notation for an inverse function is $f^{-1}(x)$.

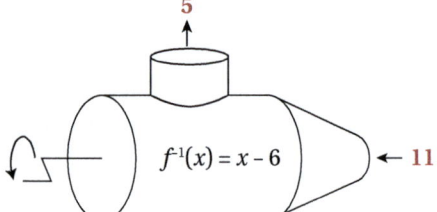

To work out the inverse of a function – particularly a more complex one – the method is:

- Write the function in the form y = the function
- Replace the 'y' with an 'x' and all the 'x's with 'y's.
- Make y the subject – you will have the inverse function.
- Write down the inverse function starting with $f^{-1}(x) =$

A few points about inverse functions which you need to know:

- The graph of an inverse function can be found by reflecting the graph of the function in the line $y = x$.
- The statement $f(5) = 11$ is exactly the same as $f^{-1}(11) = 5$.
- The domain of a function is the same as the range of its inverse, and the range of a function is the same as the domain of its inverse.
- $(f \circ f^{-1})(x) = (f^{-1} \circ f)(x) = x$ (eg: $\sqrt{x^2} = \left(\sqrt{x}\right)^2 = x$)

Example: Given that $f^{-1}(-8) = 3$, find the value of a in the function $f(x) = 10 - ax^2$.

Solution: If $f^{-1}(-8) = 3$ then it follows that $f(3) = -8$.

Thus, $10 - a \times 3^2 = -8$ giving $a = 2$.

Given the function $f(x) = \sqrt{x+2}$ for $x \geq -2$,

(a) Find the inverse function $f^{-1}(x)$

(b) Find the domain of f^{-1}

(a) $y = \sqrt{x+2}$	a) Use the steps above to find the inverse. Note that the algebra is sometimes more complicated. For example, try finding the inverse of $f(x) = \dfrac{x+3}{2x-1}$.
$x = \sqrt{y+2}$	
$x^2 = y + 2$	
$y = x^2 - 2$	
So $f^{-1}(x) = x^2 - 2$	
(b) Domain of f^{-1} is $x \geq 0$	b) Domain of inverse = range of function.

See the website for full working

Self-inverse functions: A function whose inverse is exactly the same as the original function is called *self-inverse*. In other words, $f(x) = f^{-1}(x)$. This means that for a self-inverse function $f(x)$, if $f(a) = b$ then $f(b) = a$. Take $f(x) = 10 - x$ which is a self-inverse function: for example, $f(6) = 4$ and $f(4) = 6$.

Example: Prove that $f(x) = \dfrac{x}{x-1}$ is a self-inverse function.

Solution: First we find the inverse function by writing $x = \dfrac{y}{y-1}$

Now make y the subject:

$$x = \dfrac{y}{y-1}$$

$$yx - x = y$$

$$yx - y = x$$

$$y(x - 1) = x$$

$$y = \dfrac{x}{x-1}$$

Thus $f^{-1}(x) = f(x)$ and $f(x)$ is self-inverse.

The inverse of a many-to-one function: The inverse of a one-to-one function is itself one-to-one and therefore also a function. However, the inverse of a many-to-one is a one-to-many and hence *not* a function, unless the domain of the original is restricted, hence turning it into a one-to-one function.

For example, the graph shows a quadratic with a vertex at $x = -2$.

There are no domain restrictions. However, if the inverse is to be a function, we must restrict the domain of the quadratic to remove the turning point.

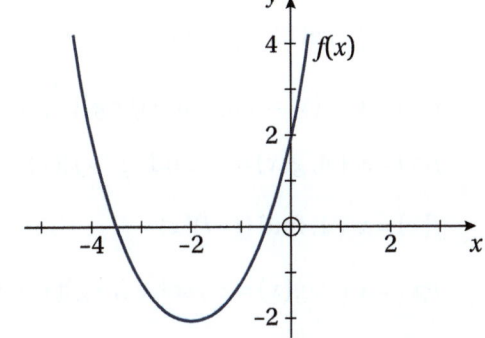

If a function has turning points it is many–one. To ensure the inverse is a function, the domain should be restricted so as not to include any turning points.

So, a suitable restriction would be $x \geq -2$. When reflected in the line $y = x$ (to get the graph of the inverse function) the result is:

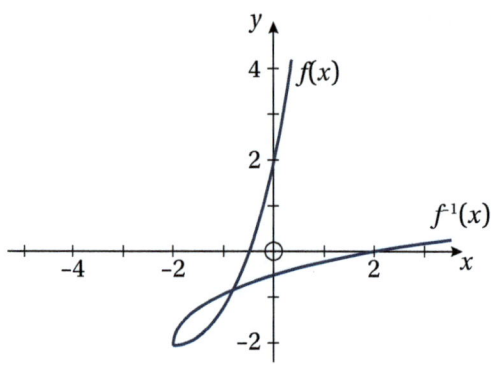

The inverse is a function since it is a one-to-one relationship.

Remember that:
$(f^{-1} \circ f)(x) = (f \circ f^{-1})(x) = x$

Composite functions: If the values of one function are input to another one, the result is a *composite function*. Given $f(x) = x^2$ and $g(x) = x - 1$ then $f(g(3)) = f(2) = 4$. This is not the same as $g(f(3)) = g(9) = 8$. It is important to understand that the functions are not being multiplied together; a number is being put through one function, then the other. This can be illustrated using the function machines again.

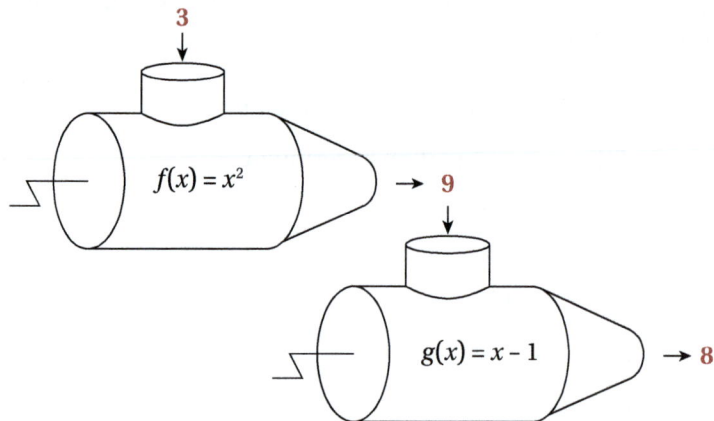

The actual notation used (to avoid multiple brackets) is $(g \circ f)(3)$. Say this as "g of f of 3" and remember that 3 is put into f first and then into g.

To find $(g \circ f)(x)$, work like this: $(g \circ f)(x) = g(f(x)) = g(x^2)$. Now, in words, function g is "subtract 1." So $g(x^2) = x^2 - 1$. Try using the same method to show that $(f \circ g)(x) = (x - 1)^2$.

Questions may also involve composites such as $f \circ f$ (put the value through function f twice), or $f^{-1} \circ g$ (put the value through function g then through the inverse of function f). Try the following examples:

$f(x) = 3x^2$, $g(x) = \frac{1}{x}$, find $(g \circ f)(x)$	$\frac{1}{3x^2}$
$f(x) = x^2$, $g(x) = \sin x$, find $(f \circ g)\left(\frac{2\pi}{3}\right)$	$\frac{3}{4}$
$f(x) = x + 4$, $g(x) = e^x$, find $(g^{-1} \circ f)(x)$	$\ln(x + 4)$
$f(x) = x + 1$, find $(f \circ f)(x)$	$x + 2$
$f(x) = x + 3$, $g(x) = 2^x$, solve $(f \circ g)(x) = (g \circ f)(x)$	-1.22 (GDC)

2. FUNCTIONS

Given that $f(x) = 4(x - 1)$ and $g(x) = \dfrac{6-x}{2}$

(a) Find g^{-1}

(b) Solve $(f \circ g^{-1})(x) = 4$

a) $y = \dfrac{6-x}{2}$

$x = \dfrac{6-y}{2}$

$2x = 6 - y$

$y = 6 - 2x$

So $g^{-1}(x) = 6 - 2x$

b) $f(g^{-1}(x)) = 4$

$f(6 - 2x) = 4$

$4(6 - 2x - 1) = 4$

$6 - 2x - 1 = 1$

$2x = 4$

$x = 2$

b) Just to reiterate – do not multiply the functions together. Put one function as the input to the second.

Odd and even functions: Consider the graphs of all the even powers of x, that is $y = x^2$, $y = x^4$ and so on. All these graphs are symmetrical around the y-axis, hence the term *even function*. The term applies to *any* function whose graph has the y-axis as a line of symmetry. Algebraically, $f(x) = f(-x)$. The graphs of the odd powers of x all have rotational symmetry about the origin, and any function whose graph has this property is called an *odd function*. The algebra in this case is $f(x) = -f(-x)$. To test if a function is even, replace x with $-x$, simplify, and see if the resulting function is the same as the one you started with. To test if it is odd, do the same thing but also put a minus sign in front of the function before simplifying.

Example: Prove that:

(a) $f(x) = \sqrt{4 - x^2}$ is even,

(b) $g(x) = x^3 - \dfrac{1}{x}$ is odd.

Solution: (a) $f(-x) = \sqrt{4 - (-x)^2} = \sqrt{4 - x^2}$

$= f(x)$

(b) $g(-x) = (-x)^3 - \dfrac{1}{(-x)}$

$= -x^3 + \dfrac{1}{x}$

$-g(-x) = -\left(-x^3 + \dfrac{1}{x}\right)$

$= x^3 - \dfrac{1}{x}$

$= g(x)$

MATHEMATICS: ANALYSIS AND APPROACHES HL

Try sketching the graphs of the following functions on your GDC and decide which are odd, which are even and which are neither:

O; N; E; O; E; N; E; O; N

$f(x) = \sin x$; $f(x) = \sin(x - \frac{\pi}{3})$; $f(x) = \cos x$; $f(x) = \tan x$; $f(x) = x^2 + 1$; $f(x) = (x+1)^2$; $f(x) = |x|$; $f(x) = |x+2| - |x-2|$; $f(x) = |x| - |x-2|$.

> Given that f is an odd function and g is an even function, determine whether the following are even, odd, or neither:
>
> (a) $f \times g$
>
> (b) $f + g$

(a) Let $h(x) = f(x) \times g(x)$

$h(-x) = f(-x) \times g(-x)$

$\quad\quad = -f(x) \times g(x)$

$-h(-x) = f(x) \times g(x)$

$\therefore fg$ is odd

(b) Let $h(x) = f(x) + g(x)$

$h(-x) = f(-x) + g(-x) = -f(x) + g(x)$

$-h(-x) = f(x) - g(x)$

$\therefore f + g$ is neither

Whenever you have to use algebraic proofs for odd or even functions, always go back to the basic definitions.

In line 3, we can replace $f(-x)$ with $-f(x)$ because f is odd, and $g(-x)$ with $g(x)$ because g is even.

The following section looks in detail at all aspects of graphs

Using a graph to answer function questions: A graph is effectively a "picture" of a function. The x-axis contains the numbers input to the function and therefore represents the domain; the y-axis contains the resulting function values, and therefore shows the range. Thus a point with coordinates (a, b) is the graphical equivalent of $f(a) = b$, and also $f^{-1}(b) = a$. For example, the graph of $y = 2^x$ is a "picture" of the function $f(x) = 2^x$, and the point $(3, 8)$ represents $f(3) = 8$. It is helpful to think of the y-axis as the "function axis".

You can answer questions about a function just from its graph, even without an equation. For example, look at this graph. Because there is a point $(5, 3)$, we know that $f(5) = 3$. What about $f^{-1}(-1)$? Work backwards from -1 on the y-axis, and we can see that $f^{-1}(-1) = -3$. Here are some more possible questions:

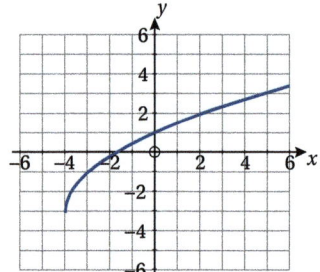

Find $(f \circ f)(-4)$.   Answer: $(f \circ f)(-4) = f(-3) = -1$

Solve $f^{-1}(a) = 0$.   Answer: $a = f(0) = 1$

What is the domain of f^{-1}?   Answer: same as the range of f, so $-3 \le x \le 3$

You may also be asked to sketch the graph of the inverse function. The easiest way to do this is to take key points, such as $(5, 3), (2, 2), (0, 1), (-3, -1), (-4, -3)$, reverse their coordinates – $(3, 5), (2, 2), (1, 0)$ etc – then plot them and join them up.

Note the difference between "draw" and "sketch". Drawing a graph will require plotting many points. A sketch shows the shape of a graph and how it relates to the axes, with a few key points marked in.

2.2 Graphs of Functions

A graph is an excellent tool for interpreting a function. From a graph we can see when the function is increasing or decreasing, what the range of the function is, where it cuts the axes and so on. Therefore it is important to be able to sketch and understand graphs of different types of functions. Remember that your calculator can be of great benefit, and you should fully understand its graphing functions (see page 45); but you must also be able to sketch graphs without a calculator.

Asymptotes: A graph such as $y = 2^x$ has a horizontal asymptote because as x gets smaller, the values of y get ever closer to 0 without ever reaching it. Some functions have graphs with vertical asymptotes which arise because division by 0 is impossible.

For example, $y = \frac{x}{2-x}$ (pictured) has a vertical asymptote at $x = 2$; as x gets closer to 2, the denominator gets closer to 0. This graph also has a horizontal asymptote at $y = -1$. Try putting x values of 10, 100 and 1000 into the function to see why.

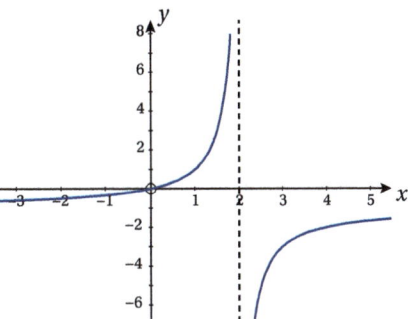

If the numerator of a rational function is one degree higher than the denominator, then the graph will have an *oblique asymptote*. Consider $f(x) = \frac{x^2 - x - 1}{x + 1}$. Algebraic division results in $f(x) = (x - 2) + \frac{1}{x+1}$ and as $x \to \infty$, $f(x) \to x - 2$ (because the fraction tends towards 0).

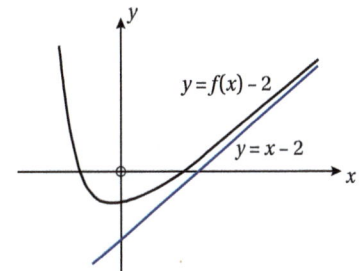

Significant features: The other main features of a graph you should be able to identify are the x-intercepts (where $y = 0$); the y-intercept (where $x = 0$); and turning points, where the gradient = 0. But note that a graph doesn't necessarily have all, or even any, of these features.

Transformations of graphs: You should be able to sketch the graphs of the basic functions $y = x^2$, $y = x^3$, $y = \frac{1}{x}$, $y = a^x$, $y = \log_b x$. The effect of simple numerical changes to these functions (involving additions, multiplications and minus signs) results in specific, simple transformations, thus extending the number of functions which can be easily sketched.

The graph transformations you need to know are:

Change to function	Transformation
$y = f(x) + a$	Translate (slide) graph upwards by a units
$y = f(x + a)$	Translate graph to the *left* by a units
$y = af(x)$	Stretch graph vertically by scale factor a
$y = f(ax)$	Stretch graph horizontally by scale factor $1/a$
$y = -f(x)$	Reflect graph in x-axis
$y = f(-x)$	Reflect graph in y-axis

For example, $y = (x - 1)^2 + 2$ will translate the graph of $y = x^2$ to the right by 1 and up by 2, that is, a translation of $\binom{1}{2}$.

Transformations in the x direction always do the opposite of what you expect!

MATHEMATICS: ANALYSIS AND APPROACHES HL

$y = -\frac{3}{x+2}$ is a composite transformation of $y = \frac{1}{x}$. To obtain the correct order of transformations, consider what order you would work out the expression if you put in a value for x. This would be:

- Add 2 to x
- Multiply the function by 3
- Change sign

The equivalent transformations are:

- Translate left 2 units
- Stretch by 3 in the y direction
- Reflect in the x-axis

> Be aware of the difference between, say, adding 2 to the x part of the function, and adding 2 to the *whole* function.

If a question asks you to sketch the transformation of a particular curve, the best way is to work out where the key points on the graph will move to, plot them, and then sketch the curve. And sometimes the transformation is not specifically mentioned.

Example: Find the turning point on the graph of $f(x) = x^2 + \frac{16}{x} - 9$ and hence write down the turning point on the graph of $f(x) = x^2 + \frac{16}{x}$.

Solution: Using differentiation, we find the turning point is when $x = 2$. Substitute this value into the function, and the turning point is (2, 3). Now to get the second function we add 9 to the first, and so the graph has been translated up 9. The turning point goes with it, and is now at (2, 12).

> On Paper 2 a GDC can be used for the first part; but because of the "hence" we must consider transformations for the second part.

The graph of $f^{-1}(x)$: Consider the graph of $y = x^2$, $x > 0$ (which represents the function $f(x) = x^2$). When $x = 3$, $y = 9$ (ie $f(3) = 9$). The graph of the inverse function is $y = \sqrt{x}$, and when $x = 9$, $y = 3$. Any point (a, b) on the graph of f becomes (b, a) on the graph of f^{-1}. So the graph of $y = f^{-1}(x)$ is always the reflection of the graph of $y = f(x)$ in the line $y = x$.

Transformations of graphs: Practice exercise

1. Find the x-intercepts, y-intercept on the graph of $y = x^2 + x - 6$. Hence find the images of these points when transformed onto the graph of $y = x^2 + x - 3$.

2. The graph of $y = 2\ln x$ is obtained from the graph of $y = \ln(x - 3)$. List the transformations required.

3. The points (0, 3) and (−2, 4) lie on the graph of $f(x)$. Write down the equivalent points on the graphs of: (a) $f(2x)$; (b) $−f(x + 1)$

4. Find the turning point on the graph of $f(x) = -x^2 + 4x - 5$. The graph of f is translated to the graph of g by the vector $\begin{pmatrix} 0 \\ a \end{pmatrix}$. Find the values of a such that $g(x) = 0$ has exactly two solutions.

5. The graph of $f(x) = \frac{1}{x}$ is transformed into the graph of $g(x)$ by a translation along the vector $\begin{pmatrix} -3 \\ 1 \end{pmatrix}$. Find an expression for $g(x)$ and write down the equations of the asymptotes of the graph of g.

Absolute value function: The graph of $|f(x)|$ is simple to sketch once you have the graph of $f(x)$: the positive parts of the graph (ie above the x-axis) remain the same, the negative parts are reflected in the x-axis.

Answers

1. (2, 0), (−3, 0), (0, −6); (2, 3), (−3, 3), (0, −3).
2. Translation $\begin{pmatrix} -3 \\ 0 \end{pmatrix}$ then stretch ×2 in the y direction.
3. (a) (0, 3), (−1, 4)
   (b) (−1, −3), (−3, −4)
4. (2, −1), $a > 1$
5. $y = \frac{1}{x+3} + 1 = \frac{x+4}{x+3}$
   $x = -3, y = 1$

The graph of $f(|x|)$ works slightly differently. Draw it as follows:

- Draw the graph of $f(x)$ but only for $x \geq 0$.
- Reflect this section in the y-axis.
- The two sections together make up the graph of $y = f(|x|)$.

Reciprocal of $f(x)$: Drawing the graph of $\frac{1}{f(x)}$ from the graph of $f(x)$ takes slightly more skill. Work as follows:

- All points with a y-coordinate of 1 or –1 remain in the same place.
- Any vertical asymptotes (discontinuities) become points on the x-axis, the x-coordinate remaining the same.
- Any points on the x-axis become discontinuities.
- The rest of the graph is completed by ensuring that values for $0 < y < 1$ become > 1, and for $-1 < y < 0$ become < -1.
- Also note that $f(x)$ and its reciprocal always have the same sign.

The graph on the right is of $y = f(x)$. The other three show the graphs of related functions as described above.

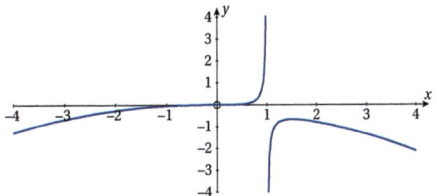

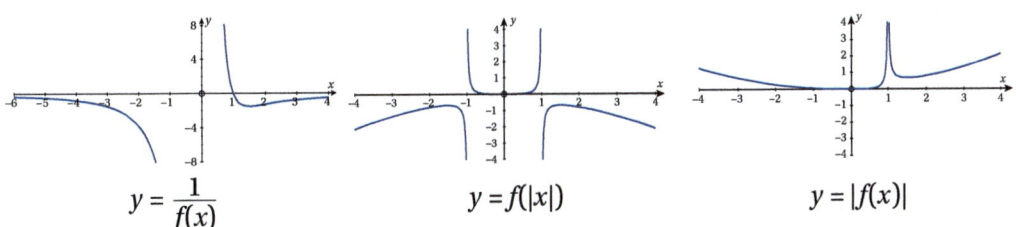

$y = \dfrac{1}{f(x)}$ \qquad $y = f(|x|)$ \qquad $y = |f(x)|$

Graphs practice: If you would like to practice the transformations above, try sketching graphs of the following functions and then, in each case, sketch the graphs of $|f(x)|$, $\dfrac{1}{f(x)}$, $f(|x|)$:

$$y = \frac{x}{2+x}, \quad y = 2e^x - 2, \quad y = \frac{x}{1-x^2}$$

Use your GDC to check your results.

Functions and Graphs with a GDC

In this section of the syllabus, perhaps more than any other, you are expected to be able to use your graphic calculator for a wide range of techniques. Apart from using your GDC for calculations, two of its most important uses (as far as exams are concerned) are for drawing graphs and for solving equations.

Function keys: You should be able to enter a wide range of functions with confidence – you don't want to spend time in the exam searching your calculator. Look at the list in the notes box – make sure you can key in each of the functions. You should also be able to work in fraction notation, and be able to enter functions such as $y = \sqrt[3]{\dfrac{x}{x-1}}$.

$\sqrt[3]{12}$

arccos 0.867

e^4

$\log_2 15$

$2 + 3i$

$4e^{i\pi}$

45

MATHEMATICS: ANALYSIS AND APPROACHES HL

Tables: GDCs have the facility to work out a table of values for a function. Having input the function in the form $y = f(x)$ you can set up a table by selecting the first x value and then the steps by which you want x to increase. In this example, the function $y = 2 - 3\sin x$ has been entered into the function editor, and then a table created starting with $x = 0$ and increasing x in steps of 30°. This can be helpful if you need to know several values, if you want to plot a graph by hand or if you're having difficulty creating the appropriate scales for a calculator plot – the table indicates the lowest and highest values of y and helps you set an appropriate window.

Drawing graphs: Three important points to remember when drawing and using GDC graphs:

> Most GDCs sold now will allow the entry of expressions in their correct format, for example 2^{x+3} instead of $2\wedge(x+3)$. It is still important to ensure the correct use of brackets.

- Make sure the function you type into the editor is actually the same as in the question. You may, for example, have to use brackets which aren't actually required on the written page. 2^{x+3}, if typed as $2 \wedge x + 3$, will calculate as $2^x + 3$. To get the correct answer you would need a bracket: $2 \wedge (x + 3)$.

- The GDC has a few standard sets of scales, but you will probably have to set up the "window" yourself in order to see the required part of the graph. You may well have to zoom into a part of the graph to see exactly what is happening. The two screenshots are of the same graph, but only the lower one shows the intersections with the x-axis.

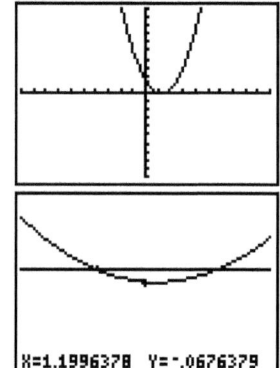

- The GDC can give you the values of key points such as intersections with the axes, points where lines intersect, turning points and so on. If you want to read off your own point, make sure you know the scales being used, ie how much each mark on the axes is worth.

Solving equations: GDCs have built in equation solvers. However, if you already have the graph of the relevant function, then it can be used to solve equations. The easiest way to do this is to ensure your equation has a 0 on the right hand side because then all you have to do is find out where the graph cuts the axis. For example, solve $x^2 - 2 = \frac{1}{x}$, $x > 0$.

First we need to rewrite this equation as $x^2 - 2 - \frac{1}{x} = 0$ and draw the graph of $y = x^2 - 2 - \frac{1}{x}$.

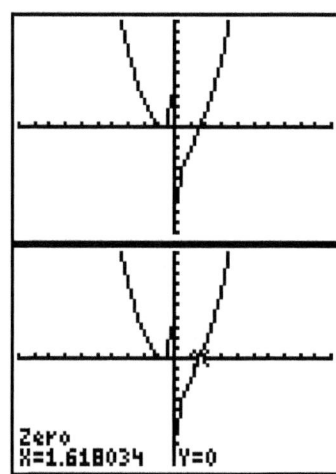

Now use the "zero" or "root" feature to find where the graph cuts the x-axis and this will be the solution to the equation, which is $x = 1.618$.

Given that $f(x) = x^3 \times 2^{-x}$, $x \geq 0$

(a) Sketch the graph of $f(x)$ showing its asymptotic behaviour.
(b) Find the co-ordinates of the maximum point, and hence state the range of $f(x)$.
(c) Draw a line on your graph to show that $f(x) = 1$ has two solutions.
(d) Find the solutions to $f(x) = 1$ giving your answers to 3 significant figures.

(a) and (c)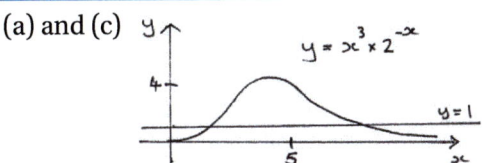

(b) Maximum = (4.33, 4.04)

Range is $0 \leq f(x) \leq 4.04$

(c) The line $y = 1$ intersects $f(x)$ at two points.

(d) $x = 1.37$ or 9.94

With a function like this we clearly need a calculator for the graph.

(a) When you sketch a graph, you must label the scales and the graph, and show at least one point on each axis to give an indication of scale. Do not go beyond the domain.

(b) and (d) You should be able to use your GDC to find turning points on a graph, and to find where two graphs intersect.

2.3 Linear Functions

In a linear function, the function increases (or decreases) at a constant rate. Its graph is a straight line.

They're called linear functions because they always give a straight line when drawn on a graph.

Example: Cost of printing programmes against number of programmes printed.

Equation: $f(x) = ax + b$ where a and b are constants.

Gradient: The *gradient* of the line is its "steepness." A gradient of 3 means that y is increasing 3 times faster than x. The gradient is calculated by choosing two points and dividing the change in y by the change in x, often remembered as $\frac{\text{rise}}{\text{run}}$.

Horizontal lines have gradient 0. Vertical lines have an infinite gradient. Lines angled from bottom left to top right have positive gradients, others have negative gradients.

Midpoint, distance between two points: The midpoint of two points can be found by calculating the x-coordinate halfway between the x-coordinates of the two points, and the same for the y-coordinate. The distance between two points is calculated using Pythagoras' Theorem.

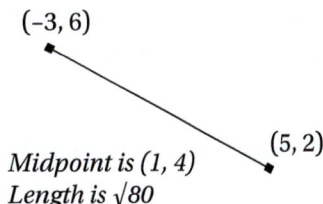

Midpoint is (1, 4)
Length is $\sqrt{80}$

Formulae

The gradient between two points (x_1, y_1) and (x_2, y_2) is $\frac{y_2 - y_1}{x_2 - x_1}$

Parallel lines: $m_1 = m_2$

Perpendicular lines: $m_1 = -\frac{1}{m_2}$

Midpoint of two points is $\left(\frac{x_1 + x_2}{2}, \frac{y_1 + y_2}{2}\right)$

Distance between two points is $\sqrt{(x_2 - x_1)^2 + (y_2 - y_1)^2}$

Although the formulae are shown in the box, this means there are a lot of formulae to remember. It is often better to draw a sketch and work from that.

MATHEMATICS: ANALYSIS AND APPROACHES HL

A useful line to remember is the one which passes through $(a, 0)$ and $(0, a)$. Its equation is $x + y = a$.

Drawing a line from its equation:

- If the equation is of the form $y = mx + c$, substitute 2 or 3 values for x and work out the corresponding y values.
- If the equation is of the form $ax + by = c$, it is easier to put x equal to 0 and work out y, then put y equal to 0 and work out x. This gives the two points where the line crosses the axes.
- To **sketch** the graph of $y = mx + c$, remember that c is the y-intercept and m is the gradient.

Working out the equation of a line from the graph:

1. Calculate the gradient.

   There are two formulae you can use:

2a. If using the first formula, replace m with the gradient, then substitute a point for x and y.

   $y = mx + c$

3a. Calculate c and then put this back into the equation.

2b. If using the second formula, replace m with the gradient then substitute the point for x_1 and y_1.

   $y - y_1 = m(x - x_1)$

3b. Rearrange and simplify to get the equation.

The points P, Q have coordinates P(3, 0), Q(-3, 7). Find the equation of the line which is perpendicular to PQ and passes through P. Give your answer in the form $ax + by + c = 0$, where a, b and c are integers.

Gradient of PQ $= \dfrac{7-0}{-3-3} = -\dfrac{7}{6}$	*Whenever you need to find the equation of a straight line your first thought should be: "I need a point, I need a gradient."*
$\therefore$ Gradient of perpendicular line to PQ $= \dfrac{6}{7}$	
Equation of line is $y - y_1 = m(x - x_1)$	
$\quad y - 0 = \dfrac{6}{7}(x - 3)$	*I prefer to use this formula when finding the equation of the line because all the numbers are substituted in one go.*
$\quad 7y = 6x - 18$	
So the line has equation $6x - 7y - 18 = 0$	

See examples in the calculus chapter from page 137 onwards.

It's unlikely you will get an exam question purely on the equations of straight lines. But you will need to use what you know to find, for example, the equation of the tangent or the normal to the point on a given curve.

2.4 Reciprocal Functions

In the reciprocal function $f(x) = \dfrac{a}{x}$, where a is a constant, the function **decreases** as the x values increase. Specifically, if an x value is multiplied by any number, the y value will be divided by the same number.

Example: The time taken to fly a fixed distance against the speed. (If the speed doubles, the time halves).

Graph: The diagram shows the graphs of two reciprocal functions. They have similar shapes. Each graph is in two sections, with the y-axis being a vertical asymptote. Since they are also self-reflections about $y = x$ this means that a reciprocal function is its own inverse. For example, $f(x) = \frac{12}{x} \Rightarrow f^{-1}(x) = \frac{12}{x}$. This is easily seen if 2 is put into the function: $\frac{12}{2} = 6$ then $\frac{12}{6} = 2$.

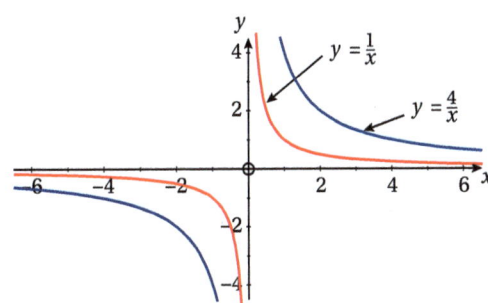

Reciprocal functions will often be found in transformation questions. For example, if the graph of $f(x) = \frac{1}{x}$ is reflected in the x-axis and then translated along the vector $\begin{pmatrix} 0 \\ -2 \end{pmatrix}$, it will be transformed into the graph of $g(x) = -\frac{1}{x} - 2$. The horizontal asymptote becomes $y = -2$, but the vertical asymptote remains $x = 0$.

And here's an example of a reciprocal function used in a composite function question:

Example: Given that $f(x) = 1 - x$ and $g(x) = \frac{1}{x}, x \neq 0$, find $h(x) = (f \circ g \circ f)(x)$.

Solution: If you put $f(x)$ into $g(x)$ you get $\frac{1}{1-x}$. Now put the new function into f (which in words is "take away from 1") and $(f \circ g \circ f)(x) = 1 - \frac{1}{1-x}, x \neq 1$.

> Now show that $h(x) = \frac{x}{x-1}$. Also find $h^{-1}(x)$ What do you notice? See website for the full working.

Rational functions: Rational functions are defined as functions which themselves have polynomial functions on both the numerator and denominator. Rational functions are often of the form $f(x) = \frac{ax+b}{cx+d}$, in which case there will always be a value of x which makes the bottom line equal to 0. This means there will always be a vertical asymptote on the graph of the function at $x = -\frac{d}{c}$. And since the values of b and d become insignificant as x approaches $\pm\infty$, there will always be a horizontal asymptote where $y = \frac{a}{c}$. However, in the HL course, you may also find rational functions of the form $f(x) = \frac{ax+b}{cx^2+dx+e}$ and $\frac{ax^2+bx+c}{dx+e}$. In the first case, there may be no value of x which makes the denominator zero, and hence no vertical asymptote – try drawing the graph of $y = \frac{2x+1}{x^2+x+1}$ on your GDC; but the x-axis will be the horizontal asymptote. In the second case there will be an oblique asymptote, and it will be necessary to divide the denominator into the numerator to find its equation.

When sketching graphs of rational functions, you will be expected to show all asymptotes and axis intercepts.

Example: Sketch the graph of the function $f(x) = \frac{2x-3}{x+1}, x \neq -1$

Solution: First, let's work out the x and y intercepts:

- When $x = 0$, $y = -\frac{3}{1} = -3$
- When $y = 0$, $0 = \frac{2x-3}{x+1} \Rightarrow x = 1.5$ (since only the numerator will be 0)

> Remember that when considering the graph of a function you can replace $f(x)$ with y.

Now the asymptotes:

- Looking at the denominator we see that $x \neq -1$ so $x = -1$ is the vertical asymptote.
- And when x takes on very large values, we can ignore the effect of the -3 and the 1, so $f(x)$ approaches $\frac{2x}{x} = 2$. Thus $y = 2$ is the horizontal asymptote.

Let's put all that information on a sketch:

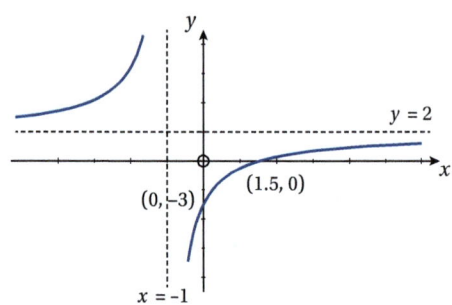

You can now deal with all rational functions in the same way. Note that it is possible for the numerator to be a constant, for example $f(x) = \frac{6}{x-3}$. In this case, there is no x intercept (since the function cannot equal zero); and the y intercept is -2. The vertical asymptote is $x = 3$; and the horizontal asymptote is $y = 0$ since the larger value x takes, the smaller the function becomes overall.

Try sketching the following functions and then checking each result by drawing the graph on your GDC. In the last example, you will have to rewrite the equation as a rational function, or you could transform the graph of $y = \frac{2}{x-1}$.

(a) $f(x) = \frac{2x+3}{2x-4}$; (b) $f(x) = \frac{2}{x+1}$; (c) $f(x) = \frac{5-2x}{x}$; (d) $f(x) = 4 + \frac{2}{x-1}$

(e) $\frac{2x}{x^2+1}$; (f) $\frac{2x}{x^2-1}$; (g) $\frac{x^2 - 2x - 1}{x-3}$

> Part (g) involves an oblique asymptote. Divide the denominator into the numerator to get $y = x + 1$.

Questions may involve functions where constants are given as letters instead of numbers. You can be pretty certain that at some stage you will be asked to find their values, often by substituting the coordinates of a point.

Let $f(x) = \frac{qx}{2x - p}$

(a) The graph of f has a vertical asymptote $x = 4$. Find the value of p.
(b) The point $(5, 10)$ lies on the graph of f. Find the value of q.
(c) Write down the equation of the horizontal asymptote to the graph of f.

(a) $2x - p = 0$ $8 - p = 0 \implies p = 8$ (b) $y = \frac{qx}{2x - 8}$ Substitute $(5, 10)$ to get $10 = \frac{5q}{2}$ So $q = 4$ (c) $f(x) = \frac{4x}{2x - 8}$ Horizontal asymptote is $y = 2$	a) For the vertical asymptote, put the bottom line equal to 0. b) When you find values of constants, always rewrite the function containing the new information. c) Again, now we know the values of p and q, we can write down the function. To find the horizontal asymptote we are interested in very large values of x, so can ignore the 8.

2.5 Quadratic Functions

Quadratic functions occur in many different situations. You should be completely familiar with the connections between the functions and their graphs, and with the methods for solving quadratic equations.

Equation: $f(x) = ax^2 + bx + c$

Graph: All quadratic graphs are parabolas, the sign of a determining "which way up." In the form shown above, we can say which way up the graph is and where the y-intercept is. For example, the graph of $y = x^2 + 3x - 4$ cuts the y-axis at $(0, -4)$ and is in the shape of a U. The graph is always symmetrical about the vertical line passing through the vertex (turning point), a fact which can often be used when answering questions about the graph. The equation of this line is $x = -\dfrac{b}{2a}$.

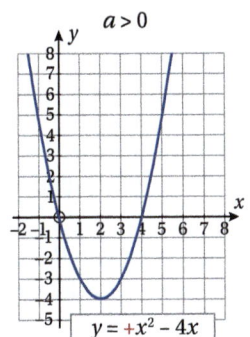
$y = +x^2 - 4x$

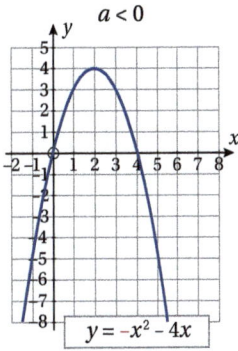
$y = -x^2 - 4x$

Factorisation: Not all quadratic expressions factorise. For those that do, the method you use depends on whether b or c equals 0.

If $c = 0$, simply take the common factor out.

$$2x^2 - 6x = 2x(x - 3)$$

If $b = 0$, you can only factorise if the quadratic is of the form $x^2 - a^2$.

$$x^2 - 25 = (x + 5)(x - 5)$$

> Expressions of the form $x^2 + a^2$ *do not factorise.*

If all three terms are present, the quadratic will factorise if you can find two numbers which add to give b and multiply to give c.

$$x^2 + 3x - 10 = (x + 5)(x - 2) \text{ because } 5 + (-2) = 3 \text{ and } 5 \times 2 = 10$$

Don't forget to look for a common factor first

$$3x^2 + 21x + 36 = 3(x^2 + 7x + 12) = 3(x + 4)(x + 3)$$

$$2x^2 - 32 = 2(x^2 - 16) = 2(x + 4)(x - 4)$$

In factorised form, the equation reveals the x-intercepts of the graph, also known as the *roots* of the quadratic equation. In the five examples above, the x-intercepts are: $(0,0)$ and $(3, 0)$; $(-5, 0)$ and $(5, 0)$; $(-5, 0)$ and $(2, 0)$; $(-4, 0)$ and $(-3, 0)$; $(-4, 0)$ and $(4, 0)$. If you are given these intercepts you can, of course, work back to the equation.

Completing the square: this method gives a third form of writing a quadratic function. The following shows the method and a corresponding example.

1. For $x^2 + bx + c$ start by writing $(x + d)^2$ where $d = b \div 2$. | $x^2 + 6x + 7$
   $(x + 3)^2 \ldots$
2. Now write down $-d^2$ | $(x + 3)^2 - 9 \ldots$
3. Write down c and simplify. | $(x + 3)^2 - 9 + 7 = (x + 3)^2 - 2$

MATHEMATICS: ANALYSIS AND APPROACHES HL

For quadratics where $a \neq 1$, start by taking a out as a common factor. Forget about it whilst completing the square. Multiply it back at the end.

$$2x^2 - 6x - 4$$
$$= 2(x^2 - 3x - 2)$$
$$= 2((x - 1.5)^2 - 2.25 - 2)$$
$$= 2((x - 1.5)^2 - 4.25)$$
$$= 2(x - 1.5)^2 - 8.5$$

In this form, the function can be seen to be a transformation of $y = x^2$. In the first example above, the transformation is a translation of $\begin{pmatrix} -3 \\ -2 \end{pmatrix}$. Since the vertex of $y = x^2$ is $(0, 0)$, the vertex of the new quadratic will be $(-3, -2)$. In general, the completed square form is always $f(x) = a(x - h)^2 + k$ and this gives a vertex of (h, k).

> When an exam question says write in the form $a(x - h)^2 + k$, you are being asked to complete the square.

Example: Express $f(x) = x^2 - 4x + 9$ in the form $f(x) = (x - h)^2 + k$. Hence, or otherwise, write down the coordinates of the vertex of the parabola with equation $y = x^2 - 4x + 9$.

Solution: $f(x) = (x - 2)^2 - 2^2 + 9 = (x - 2)^2 + 5$. The vertex is $(2, 5)$. For the "otherwise", we could have found the x-coordinate of the vertex using differentiation, and then substituted to find the y-coordinate. We could also have found the x-coordinate using the fact that the line of symmetry is given by the formula $x = -\dfrac{b}{2a} = -\dfrac{-4}{2} = 2$.

Let $f(x) = a(x - 3)^2 + 6$.

(a) Write down the coordinates of the vertex of the graph of f.
(b) Given that $f(7) = -18$, find the value of a.
(c) Hence find the y-intercept of the graph of f.

(a) $(3, 6)$ (b) $-18 = a(7 - 3)^2 + 6$ $-18 = 16a + 6$ $a = -1.5$ (c) So $f(x) = -1.5(x - 3)^2 + 6$ When $x = 0$, $y = -1.5 \times 9 + 6 = -7.5$ y-intercept is $(0, -7.5)$	a) The function is in completed square form, so we can write down the vertex with no further working. b) $f(7) = -18$ is exactly the same as saying that when $x = 7$, $y = -18$. Substitute into the equation. c) The y-intercept is found by putting $x = 0$. But first, having found a, write out the equation.

2.6 Solving Quadratic Equations

> $x^2 = 25$
> $x = \pm 5$

Except for the simplest form of quadratic equation shown in the margin, the first move is *always collect together terms on the left with 0 on the right.*

> Example: $2x^2 - 4x = x^2 - 3$
> $x^2 - 4x + 3 = 0$
> $(x - 3)(x - 1) = 0$
> $x = 3$ or 1

Factorisation: If the quadratic expression factorises, this is the simplest method of solution. Make sure you understand the connection between the factors and the x-intercepts (see previous section) since questions can link the equation to the graph.

2. FUNCTIONS

Formula: *All* quadratic equations (if they have solutions) can be solved using the formula, although it is most useful when the expression does not factorise. The solution of $ax^2 + bx + c = 0$ is: $x = \dfrac{-b \pm \sqrt{b^2 - 4ac}}{2a}$. It is the $\pm$ which leads to the two possible solutions.

Your GDC will also have a quadratic equation solver you can use.

Be careful to substitute correctly, particularly when there are minus signs around. Follow the second example in the notes box carefully.

Example: $2x^2 - 4x = x + 2$
$2x^2 - 5x - 2 = 0$
$x = \dfrac{-(-5) \pm \sqrt{(-5)^2 - 4 \times 2 \times (-2)}}{2 \times 2}$
$x = \dfrac{5 \pm \sqrt{41}}{4}$
$\therefore x = 2.851$ or -0.351

> The graph of $f(x) = x^2 + cx + d$ has a line of symmetry $x = -2.5$. The distance between the two zeros (roots) is 7. Find the value of the two zeros, and hence the values of c and d.

The two zeros are equidistant from the line of symmetry, so will be at 3.5 either side of –2.5. So the zeros are at $x = 1$ and $x = -6$. The quadratic is $f(x) = (x-1)(x+6)$ $= x^2 + 5x - 6$ and so $c = 5$, $d = -6$.	*The important thing to understand in this question is the connection between the algebra and the geometry. A question could work from factors to roots to line of symmetry/ turning point; this one does all of that in reverse.*

Discriminant: The solutions of a quadratic equation are the points where the graph of the quadratic crosses the x-axis: the diagrams show that this could mean 0, 1 or 2 solutions. In the formula, *if the value of $b^2 - 4ac < 0$*, we would be trying the find the square root of a negative number – not possible, so no solutions. *If the value = 0* there will be 1 solution, and *if it's > 0*, there are two solutions. Because it discriminates between the number of solutions, $b^2 - 4ac$ is known as the *discriminant*, and can be represented by the symbol Δ (Greek delta). So any question which asks you about the number of solutions of a quadratic equation will in fact be about the discriminant.

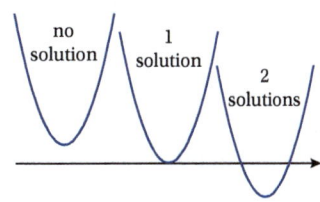

Questions may use the wording "two distinct roots" for 2 solutions, "two equal roots" for 1 solution.

One point to note: when there is one solution, it is often called a "repeated root" or "two equal roots." This is because it always arises from an equation such as $(x - 4)^2 = 0$, the solution $x = 4$ effectively being repeated.

This example shows the typical use of the discriminant in a question.

Example: Find the range of values of p for which $x^2 - px + (p + 3) > 0$ for real x.

Solution: This is a quadratic above the x-axis and so has no solutions. The discriminant will be less than zero.

$$(-p)^2 - 4(p + 3) < 0$$
$$p^2 - 4p - 12 < 0$$
$$(p - 6)(p + 2) < 0$$

This is a quadratic inequality (see page 59) with solution $-2 < p < 6$.

MATHEMATICS: ANALYSIS AND APPROACHES HL

> Find the possible values of k such that the graphs of $y = x^2 + k$ and $y = kx - 3$ have two intersections.

Graphs intersect when $x^2 + k = kx - 3$	To find the points where two graphs intersect, put the two equations equal to each other. We then see that we have a quadratic equation, so write it in the form $ax^2 + bx + c = 0$.
$x^2 - kx + (k + 3) = 0$	
$\Delta = (-k)^2 - 4(k + 3) = k^2 - 4k - 12$	
Two solutions when $\Delta > 0$, so:	Now we aren't asked to find the points, but the values of k which give us two solutions – hence the need to find the discriminant. This leads to a quadratic inequality – make sure you know how to solve these.
$k^2 - 4k - 12 > 0$	
$(k + 6)(k - 2) > 0$	
$k < -6$ or $k > 2$	

As I have mentioned before, some quadratic equations are dressed up in disguise – be on the lookout for such situations. For example, take the equation $4^{x-1} = 2^x + 8$. First step, replace 4 by 2^2, giving $2^{2x-2} = 2^x + 8$; then, remembering that 2^{2x} is the same as $(2^x)^2$, the equation becomes $(2^x)^2 - 4 \times 2^x - 32 = 0$ which is of the form $y^2 - 4y - 32 = 0$. $y = -4$ or 8, and so $2x = 8$ giving $x = 3$. (Why can't $2x = -4$?)

Quadratics: Practice Exercise

Answers

1. (a) -3, 5; (b) 1, 4;
   (c) -1.5; 4, (d) -1, -7
2. (a) Min (2, -2);
   (b) Min (1, 4);
   (c) Max (1.5, 7.25)
3. (a) -2.22, 0.225;
   (b) -1.37, 0.366
4. $p < -1$ or $p > 1$
5. $2x^2 - x - 66 = 0$,
   6 cm × 4 cm
6. $k = 20$, (2.5, 0), (2, 1) and (3, 1)
7. $a = 2$, $x = 2$, (2, 8)

Quadratics are extremely important in this course. Many algebraic situations tend to resolve to a quadratic equation, and you could be expected to use any of the methods above to solve them. It is inevitable that you will find a number of questions involving quadratics in both papers, so here's the chance for some practice in the various techniques you will need.

1. Solve these quadratics using factorisation, rearranging into the form $ax^2 + bx + c = 0$ as necessary:

   (a) $x^2 - x + 2 = x + 17$

   (b) $x = 5 - \frac{4}{x}$

   (c) $2x^2 - 5x - 12 = 0$

   (d) $20x^2 + 160x + 140 = 0$

2. Write each of the following in the form $a(x - h)^2 + k$, and state the turning point on the associated graph, and whether it is a maximum or minimum:

   (a) $x^2 - 4x + 6$

   (b) $2x^2 - 4x + 6$

   (c) $5 + 3x - x^2$

3. Solve, giving the answers the form $a + b\sqrt{c}$, where $a, b, c \in \mathbb{Q}$:

   (a) $2x^2 + 4x - 1 = 0$

   (b) $(x - 1)^2 = 6 - (x + 2)^2$

4. The equation $x^2 - 2px + 1 = 0$ has two distinct roots. Find the set of possible values of p.

5. The area of the shape on the right is 33 cm².

   Set up an equation in x and solve to find the dimensions of the rectangle.

   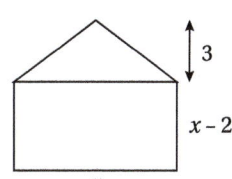

6. The function $f(x) = 4x^2 - kx + 25$ where $k > 0$ has its vertex on the x-axis. Find the value of k, the coordinates of the vertex, and the point of intersection of the graph of f with the line $y = 1$.

7. The graph shows the function $f(x) = 8x - ax^2$. The x-intercepts are at $(0, 0)$ and $(4, 0)$. Find the value of a, the equation of the line of symmetry, and the coordinates of the vertex.

   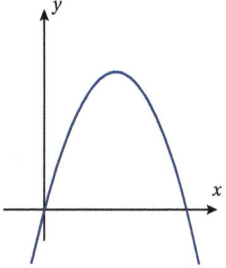

2.7 Polynomial Functions

Polynomials are of the form $a_n x^n + a_{n-1} x^{n-1} + \ldots + a_1 x + a_0$ and are easy to deal with. Differentiation and integration, for example, are a piece of cake. When polynomials factorise this leads to the easy solution of polynomial equations.

The factor theorem: There is a set routine for factorising a quadratic. For any higher degree polynomial, we can test possible factors using the factor theorem. This states that, for a polynomial function $f(x)$, $(x - a)$ is a factor if $f(a) = 0$.

The degree of a polynomial is the highest power of x.

Example: Show that $(x - 2)$ is a factor of $f(x) = x^3 - 3x^2 - 2x + 8$.

Solution: $f(2) = 2^3 - 3 \times 2^2 - 2 \times 2 + 8 = 0$, so $(x - 2)$ is a factor

If we were testing $x + 2$, then we would calculate $f(-2)$.

Once we know a factor we can divide it into the polynomial to find the other factors. Compare this with numbers – suppose we want the factors of 30 and know that 5 is one of them:

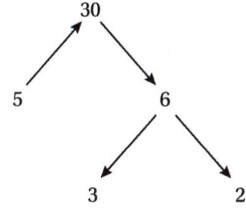

Thus, the prime factors of 30 are 2, 3 and 5.

How do we do polynomial division? You may have been taught methods such as long division or synthetic division, but comparing coefficients (which we might call *by inspection*) is quick and simple. Using the example above, we can see that we are dividing a linear factor into a cubic polynomial: the other factor must be quadratic.

$$(x - 2)(ax^2 + bx + c) = x^3 - 3x^2 - 2x + 8$$

Stage 1: First look at the coefficient of the highest power, x^3. This tells us that a must be 1. Next, look at the constant term, and this tells us that $c = -4$.

Stage 2: Now look at any of the remaining terms, for example $-2x$. This comes from the sum of $-2bx$ and cx. So $-2b + c = -2$, and we already know that $c = -4$. Thus $b = -1$.

Stage 3: Now put all the information together:

$(x - 2)(x^2 - x - 4) = x^3 - 3x^2 - 2x + 8$

The quadratic will not yield any rational factors so that's as far as we can go in this case.

The remainder theorem: If $(x - a)$ is not a factor of $f(x)$ then it will leave a remainder when $f(x)$ is divided by $(x - a)$. This remainder can be calculated as $f(a)$. So the remainder when $x^3 - 3x + 1$ is divided by $(x + 2)$ is $(-2)^3 - 3 \times (-2) + 1 = -1$.

> Thus the factor theorem is really just a special case of the remainder theorem - a factor leads to a remainder of 0.

Solution of polynomial equations: How many real solutions does $f(x) = 0$ have if $f(x)$ is a cubic function? If the graph has two turning points, there are three possibilities:

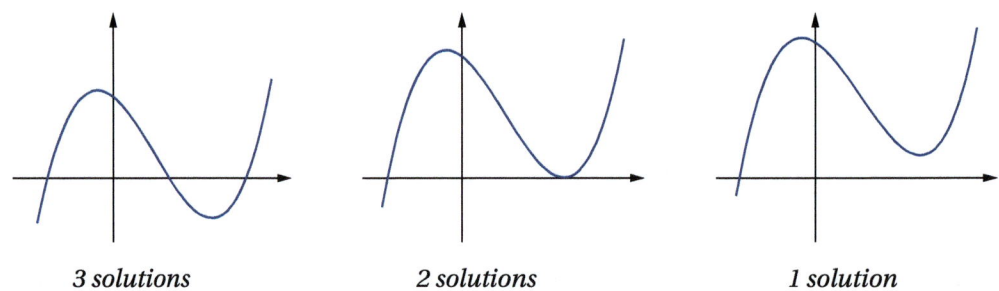

3 solutions   2 solutions   1 solution

These geometric configurations have algebraic equivalents which can be seen when the polynomial has been fully factorised:

- 3 solutions from three distinct linear factors

    eg: $(x - 4)(x - 2)(x + 1) = 0$ has solutions 4, 2 and -1.

- 2 solutions from one linear factor and one "repeated" factor

    eg: $(x + 3)(x - 1)^2 = 0$ has solutions -3 and 1. The repeated factor is the "touch" on the axis as opposed to the "cut."

- 1 solution from one linear one irreducible quadratic factor

    eg: $(x + 2)(x^2 + x + 1) = 0$ has solution $x = -2$

If we allow complex solutions then the 1 real solution becomes 1 real and a complex conjugate pair (from the quadratic).

Example: Show that $x = 1$ is a solution of $2x^3 - x^2 + 8 = 7x + 2$, and find any further solutions.

Solution: We can first rewrite the equation as $2x^3 - x^2 - 7x + 6 = 0$.

Substitute 1 in the LHS to get $2 - 1 - 7 + 6 = 0$.

Thus $x = 1$ is a solution. It follows that $x - 1$ is a factor.

$(x - 1)(ax^2 + bx + c) = 2x^3 - x^2 - 7x + 6$

$a = 2$ and $c = -6$

Comparing coefficients for x, $c - b = -7 \Rightarrow b = 1$

Thus $2x^3 - x^2 - 7x + 6 = (x - 1)(2x^2 + x - 6) = (x - 1)(2x - 3)(x + 2)$, and the two remaining solutions are $x = 1.5$ and $x = -2$.

> The function $f(x) = x^3 + ax^2 - 2x + b$ has $(x + 1)$ as a factor, and leaves a remainder of -3 when divided by $x - 2$. Find the values of a and b.

By the factor theorem, $f(-1) = 0$	*A question which has two unknowns in an algebraic expression is almost certainly going to ask you to find their values. Expect simultaneous equations – and look for two "facts" which will produce those equations.*
$\quad (-1) + a + 2 + b = 0$	
$\quad a + b = -1$	
By the remainder theorem, $f(2) = -3$	
$\quad 8 + 4a - 4 + b = -3$	
$\quad 4a + b = -7$	
Subtracting, $3a = -6 \Rightarrow a = -2$	
Substituting, $-2 + b = -1 \Rightarrow b = 1$	

2.8 Sum and Product of Roots

We know how to find the roots of a quadratic equation (the same as the zeroes of the relevant quadratic function). But, given the roots, how can we find an equation which has those roots?

For an equation, roots and solutions have the same meaning.

Reversing the factorisation process, we can work like this:

- A quadratic function has roots $x = -2$ and $x = 5$
- In factorised form, the equation must be $(x + 2)(x - 5) = 0$
- So a quadratic equation with roots -2 and 5 is $x^2 - 3x - 10$

Note that I have said 'a' quadratic equation, not 'the' quadratic equation. This is because, in general, $a(x + 2)(x - 5) = 0$ has roots -2 and 5. But unless we are given more information, such as a point on the curve, we cannot know what a is.

An important fact is that for the equation $ax^2 + bx + c = 0$ the sum of its roots is $-\frac{b}{a}$ and the product of its roots is $\frac{c}{a}$.

What would the equation be if we were told that the function has zeroes -2 and 5, and also passes through the point $(-3, -24)$? You should find that a has value -3.

In the example above, by putting $a = 1$, we get $b = -(-2 + 5) = -3$ and $c = -2 \times 5 = -10$ and so can write down the quadratic immediately. This is particularly useful when dealing with complex numbers.

Example: Find a quadratic whose roots are $(2 - 3i)$ and $(2 + 3i)$.

Solution: The sum of the roots is $(2 - 3i) + (2 + 3i) = 4$

The product of the roots is $(2 + 3i)(2 - 3i) = 4 - 9i^2 = 13$.

Thus a quadratic equation with those roots is $z^2 - 4z + 13 = 0$, having put $a = 1$.

MATHEMATICS: ANALYSIS AND APPROACHES HL

> A polynomial with real coefficients $p(x) = ax^3 + bx^2 + cx + d$ has two of its roots -1 and $1 + \frac{1}{2}i$.
>
> (a) Write down the third root.
> (b) Given that $p(1) = 4$, find the values of a, b, c, d.

(a) $1 - \frac{1}{2}i$

(b) The sum of the two complex roots is 2

The product of the two complex roots is $\frac{5}{4}$

The quadratic factor is therefore $x^2 - 2x + \frac{5}{4}$

So $p(x) = a(x + 1)(x^2 - 2x + \frac{5}{4})$

$\quad = a(x^3 - x^2 - \frac{3}{4}x + \frac{5}{4})$

$p(1) = 4 \Rightarrow 4 = a(1 - 1 - \frac{3}{4} + \frac{5}{4})$

$4 = \frac{1}{2}a$

$a = 8$

Thus $p(x) = 8x^3 - 8x^2 - 6x + 10$

(a) Complex roots always occur in conjugate pairs.

(b) Once you know the factors of a polynomial, you can pretty well find out everything about it. So use sum and product of the roots of the quadratic factor to find the factor, and then work from there.

Sometimes the focus shifts to the sum and products themselves. For example, if α and β are the roots of $x^2 - 3x - 5 = 0$, can we find the quadratic equation whose roots are $\alpha - 3$ and $\beta - 3$?

From the given equation we know that $\alpha + \beta = 3$ and $\alpha\beta = -5$. The new equation has roots with sum $(\alpha - 3) + (\beta - 3) = \alpha + \beta - 6 = -3$

And the product will be $(\alpha - 3)(\beta - 3) = \alpha\beta - 3(\alpha + \beta) + 9 = -5$

Thus the new equation is $x^2 + 3x - 5 = 0$

Now try finding quadratic equations whose roots are:

(a) $\frac{1}{\alpha + 1}, \frac{1}{\beta + 1}$ (b) $\frac{1}{\alpha^2}, \frac{1}{\beta^2}$

The answers are:
(a) $x^2 + 5x - 1 = 0$ and
(b) $25x^2 - 19x + 1 = 0$.

Full worked answers are on the website

Sum and product for all polynomials: The results for quadratics can be extended to any polynomial with degree n. If the coefficient of the degree n term is a, the coefficient of the degree $n - 1$ term is b, and the constant term is k, then the sum of the roots will be $-\frac{b}{a}$ and the product will be $\frac{k}{a}$ for even n and $-\frac{k}{a}$ for odd n. Eg, for $2x^4 + 3x^3 - x^2 + 4x + 5 = 0$, the sum of the roots is $-\frac{3}{2}$ and the product of the roots is $\frac{5}{2}$.

Sum and Product of Roots: Practice Exercise

Answers
1. $x^2 - 6x + 25 = 0$
2. $2, \frac{5}{3}$
4. $-3, \frac{3}{2}, -\frac{11}{4}$;
   $2x^2 - 11x + 8 = 0$

1. Find a quadratic equation with roots $3 + 4i$, $3 - 4i$.

2. What are the sum and the product of the roots of the equation $3x^4 - 6x^3 - 2x + 5 = 0$?

3. Given α and β are the roots of the equation $x^2 - bx + c = 0$, show that $(\alpha^2 + 1)(\beta^2 + 1) = (c - 1)^2 + b^2$.

   You will need to multiply out the left hand side of the expression, and use the fact that $p^2 + q^2 = (p + q)^2 - 2pq$.

4. If the quadratic equation $2x^2 - x - 7 = 2x - 1$ has roots α and β, find the values of $\alpha\beta$, $\alpha + \beta$, $\frac{\alpha}{\beta} + \frac{\beta}{\alpha}$. Also find a quadratic equation whose roots are $\alpha + 2$ and $\beta + 2$.

2.9 Inequalities

Linear inequalities: Linear inequalities are solved in exactly the same way as equations except for one important difference: when multiplying or dividing by a negative number, the inequality sign must be turned round.

eg: $8 - 2x > 4$

$-2x > -4$

$x < 2$

> ...or move the x term to the other side:
> $8 - 2x > 4$
> $8 > 4 + 2x$
> $4 > 2x$
> $2 > x$ (ie $x < 2$ as before).

Quadratic and cubic inequalities: Remember that if $x^2 < 4$, then it follows that $-2 < x < 2$. And if $x^2 > 4$, $x > 2$ or $x < -2$. Quadratic inequalities will always result in ranges such as this. In more complicated ones, factorise first:

eg: $x^2 - 3x - 18 < 0$

$(x - 6)(x + 3) < 0$

So $-3 < x < 6$ (it is helpful to consider the graph).

> You can only use the "double inequality" when x lies in a single range. When x lies in one of two possible ranges, use two inequalities.

A cubic in factorised form can be treated in exactly the same way; again, you will find a sketch graph useful. $(x + 2)(2x - 1)(x - 4) > 0$ has solutions $-2 < x < \frac{1}{2}$ or $x > 4$.

Solving $f(x) > g(x)$: You can solve more complex inequalities using a graphical method on your GDC.

GDC method: One way is to draw the graphs of both functions and then identify the domain values where one graph is above the other. It is generally easier, though, to draw the graph of $f(x) - g(x)$ and find where the values are greater than 0.

For example, solve $x^2 > e^{0.5x}$. The GDC display on the right shows the graph of $y = x^2 - e^{0.5x}$ (after some judicious changes of scale). The GDC gives the zeroes as -0.816, 1.430 and 8.613. So the solution to the inequality is $x < -0.816$ or $1.430 < x < 8.613$.

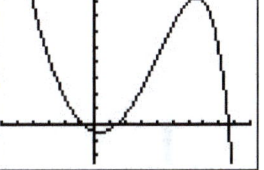

Rational inequalities will produce asymptotes, and it is just a matter of extending the process to a more complex graph. For example, to solve $\frac{2}{x-2} > \frac{1}{x-3}$, we first need to draw the graph of $y = \frac{2}{x-2} - \frac{1}{x-3}$. The first attempt (right) isn't very helpful, and shows the importance of choosing sensible scales. (Why is it obvious from the equation that there are asymptotes?) Try for yourself, and you should find that the only zero is at $x = 4$. Taking the asymptotes into account, the solution is $2 < x < 3$, or $x > 4$.

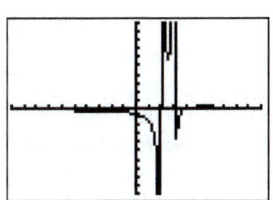

This worked example is a bit of a sneaky one:

> (a) Write $f(x) = x^2 - x + 1$ in the form $(x - h)^2 + k$.
> (b) Hence solve the inequality $\frac{x-2}{x^2-x+1} < 0$.

(a) $x^2 - x + 1 = (x - \frac{1}{2})^2 + \frac{3}{4}$ (b) $(x - \frac{1}{2})^2 + \frac{3}{4} > 0$ for all values of x Thus $\frac{x-2}{x^2-x+1}$ is negative when $x - 2$ is negative. Hence $x < 2$	(b) A square term is always positive – so adding a number on will make it even more positive. In which case, the sign of the fraction only depends on the sign of $x - 2$. Try solving by GDC for practice.

If the inequality involves a modulus sign, a graphical method is always best!

MATHEMATICS: ANALYSIS AND APPROACHES HL

2.10 Exponential and Logarithmic Functions

Equations: $f(x) = a^x$ where $a > 0$

$f(x) = \log_a x$ where $x > 0$

Notice the domains. Remember, it is not possible to take the logarithm of a negative number.

Graphs: In the same way that all quadratics have the same shape, graphs of $y = a^x$ have the same shape for different values of a, and all pass through (0, 1). Since $\log_a x$ is the inverse of a^x their graphs are reflections of each other in the line $y = x$, and the graph of $y = \log_a x$ passes through (1, 0) for all a.

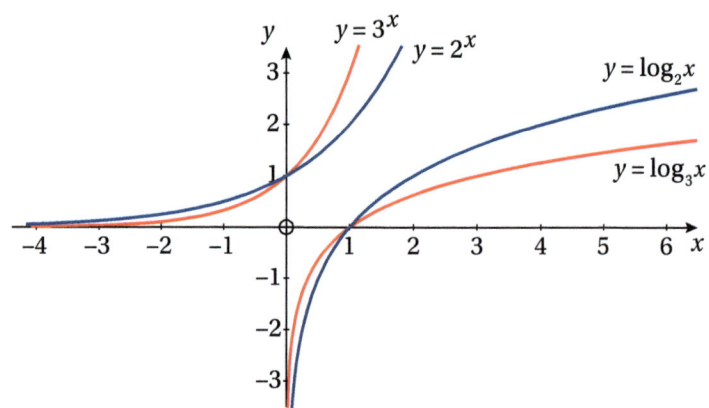

Questions draw on your knowledge of the laws of indices and logarithms. You must be familiar with the following rules:

- If $a^x = b$ then $x = \log_a b$ (eg: $2^3 = 8$, so $\log_2 8 = 3$)
- $x = \log_a a^x$ This is similar to $x = \sqrt{x^2}$
- $x = a^{\log_a x}$ This is similar to $x = (\sqrt{x})^2$

e^x and $\ln x$: The number e is, like π, given a letter because it is irrational and hence impossible to write accurately using decimals. It is approximately 2.718. The functions e^x and e^{-x} are important because they are used to model situations where the rate of growth or decay of the quantity x is dependent on the value of x at any time. Typical applications are population growth and radioactive decay. The inverse of e^x is $\ln x$, short for $\log_e x$.

> Using the rules in the previous paragraph, remember that $e^{\ln x}$ and $\ln e^x$ both simplify to x. For example, $e^{2\ln x} = e^{\ln x^2} = x^2$

Example: What are the domain and range of $f(x) = \sqrt{\ln(x - 3)}$?

Solution: Logarithms of numbers less than 1 are negative and so do not have square roots. However, $\sqrt{0} = 0$ and $\ln 1 = 0$, so $f(4) = 0$ and the domain is therefore $x \geq 4$. The range is $f(x) \geq 0$.

2. FUNCTIONS

A group of ten monkeys is introduced to a zoo. After t years the number of monkeys, N, is modelled by $N = 10e^{0.3t}$.

(a) How many monkeys are there after 3 years?

(b) How many complete months will it take for the number of monkeys to reach 50?

(a) $N = 10\,e^{0.3 \times 3} = 24.596...$ ∴ There are 24 monkeys (b) $50 = 10\,e^{0.3t}$ $5 = e^{0.3t}$ $\ln 5 = \ln e^{0.3t}$ $\ln 5 = 0.3t$ $t = 5.36...$ years $= 64.38$ months ∴ There will be 50 monkeys after 65 months	*An important point to note in the answers to this question is that they are not rounded in the normal way. In part (a) there are only 24.596 monkeys after three years, so the 25th monkey hasn't appeared yet! The answer would still be 24 even if the calculation came to 24.99.* *And in part (b), although 64.38 rounds to give 64, there won't be 50 monkeys after 64 months. In terms of complete months, the answer is 65.*

You will find plenty more practice with logarithmic and exponential functions in the Calculus chapter starting on page 137.

Functions: Practice Exercise

Here are some slightly more testing questions covering many of the topics in this chapter.

1. Let $f(x) = \dfrac{1-x}{1+x}$, $x \neq -1$ and $g(x) = \sqrt{x+1}$, $x > -1$. Find the set of values of x for which $f'(x) \leq f(x) \leq g(x)$.

   This is definitely one for the GDC. Draw the three graphs, adjust the window (note the domain restriction), identify $f(x)$ and use graph intersections to find the solution. It is possible to draw $f'(x)$ without working it out – do an internet search to find the method.

2. Given that $f(x) = \sqrt{x}$ and $g(x) = (x+5)$ express in terms of f and g:

   (a) (i) $\sqrt{x+5}$; (ii) $x+10$; (iii) x^2+5

   (b) Also express $h(x) = x^2 + 10x + 20$ in terms of f and g by first writing $h(x)$ in the form $(x-a)^2 + b$.

3. Find the equation of the curve obtained when the graph of $y = \dfrac{x^2-1}{x+2}$ is first reflected in the y-axis and then translated along the vector $\begin{pmatrix} -1 \\ 2 \end{pmatrix}$. Give your answer as a single fraction in its simplest form.

4. The remainder when $3x^3 - 6x^2 + ax - 1$ is divided by $(x+1)$ is equal to the remainder obtained when the same expression is divided by $(x-3)$. Find the value of a.

5. (a) In each of the following, determine the range of values of x for which:

   (i) $1 - 2x > 0$
   (ii) $x^2 - x - 2 > 0$
   (iii) $\dfrac{x^2 - x - 2}{1 - 2x} > 0$

   You should be able to solve the inequalities in parts (a)(i) and (a)(ii) both with and without a GDC. Part (a)(iii) requires a GDC.

Answers

1. $0 \leq x \leq \sqrt{3}$
2. (a) gf, gg, gf^{-1}; (b) $g^{-1}f^{-1}g$
3. $y = \dfrac{x^2 + 2}{1 - x}$
4. -9
5. (a) (i) $x < \tfrac{1}{2}$
   (ii) $x < -1$, $x > 2$
   (iii) $x < -1$, $\tfrac{1}{2} < x < 2$
   (b) $-4 < x < 0$
6. (a) $(1.5, 0)$, $(-1, 0)$, $(0, 0.75)$; $x = 2$, $x = -2$, $y = 2$
   (b) $(1.5, 0)$, $(-1, 0)$, $(0, -0.75)$; $y = 2$
7. $5, -\tfrac{5}{2}, -\tfrac{1}{2}$
8. $\dfrac{-kx-5}{x-3}$; -3; $2x - 1$

(b) Find the set of real values of x for which $|x - 2| > 2|x + 1|$

6. By first finding the axis intercepts, and then any asymptotes, sketch the graph of:

    (a) $y = \dfrac{(2x - 3)(x + 1)}{x^2 - 4}$ (b) $y = \dfrac{(2x - 3)(x + 1)}{x^2 + 4}$.

7. If the three roots of $2x^3 + 5x^2 - 4x - 10 = 0$ are α, β, γ, find the values of:

    (a) $\alpha\beta\gamma$ (b) $\alpha + \beta + \gamma$ (c) $\dfrac{1}{\alpha\beta} + \dfrac{1}{\beta\gamma} + \dfrac{1}{\alpha\gamma}$

8. Let $f(x) = \dfrac{3x - 5}{x + k}$, $x \neq -k$, $k \in \mathbb{Z}$.

    (a) Find an expression for $f^{-1}(x)$.

    (b) Find the value of k such that f is a self-inverse function.

    (c) Find $g(x)$ such that $(f \circ g)(x) = \dfrac{3x - 4}{x - 2}$.

Functions: Long Answer Questions

There follows a selection of section B style questions related to the Functions topic, although some of them require knowledge and techniques from other areas of the syllabus. The answers are given here, but full working may be found on the website.

2. FUNCTIONS

1. $f(x) = x^3 - 6x^2 + kx + 10$

   (a) Find the roots of the cubic equation $f(x) = 0$ given that they are all real and form the consecutive terms of an arithmetic sequence.

   (b) Hence find the value of k.

   (c) Using the value of k found in part (b), solve the equation $f''(x) = f'(x)$

   Answers:

   (a) $-1, 2, 5$

   (b) $k = 3$

   (c) $x = 1$ or 5

2. The function $f(x)$ is defined by $f(x) = (x-2)^2 + 1$ for $x \geq k$.

   (a) State the minimum value of k such that the inverse of f is also a function.

   (b) Find $f^{-1}(x)$.

   (c) State the domain and range of f^{-1}.

   (d) The graph of $g(x)$ is obtained by a reflection of the graph of f in the x-axis, then a translation along the vector $\begin{pmatrix} 0 \\ 4 \end{pmatrix}$

      (i) Find $g(x)$

      (ii) Solve $f(x) = g(x)$ given the value of k found in part (a).

   (e) Find $h(x)$ in the form $ax + b$, $a > 0$, such that $(f \circ h)(x) = 25x^2 + 20x + 5$

   Answers:

   (a) 2

   (b) $2 + \sqrt{x-1}$

   (c) $x \geq 1, f^{-1}(x) \geq 2$

   (d) (i) $g(x) = 3 - (x-2)^2$

      (ii) $x = 3$

   (e) $5x + 4$

3. Part of the graph of $f(x) = \dfrac{ax - c}{4 - bx}$ is shown right. The asymptotes are $x = 2$ and $y = -2.5$.

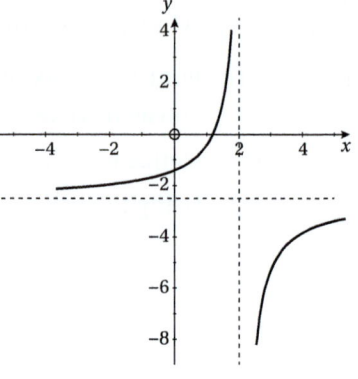

(a) Find the values of a and b.
(b) Find the value of c given that the x intercept is $(1.2, 0)$.
(c) Hence find the y intercept.
(d) Sketch the graphs of $y = |f(x)|$ and $y = f(|x|)$
(e) For $x > 0$, state the values of x for which $|f(x)| = f(|x|)$

Answers:

(a) $a = 5, b = 2$
(b) $c = 6$
(c) $(0, -1.5)$

(d)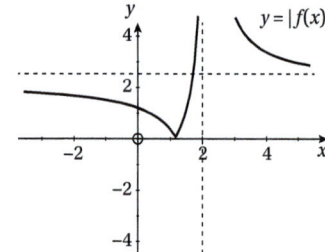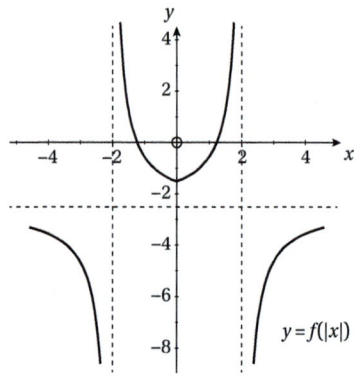

(e) $1.2 \leq x < 2$

4. Let $f(x) = \frac{1}{x} + 3$, $x \neq 0$ and $g(x) = \ln x$, $x > 0$

   (a) The graph of f is reflected in the y-axis and then translated along the vector $\begin{pmatrix} 2 \\ -1 \end{pmatrix}$. What is the equation of the transformed graph?

   (b) Find $h(x) = (f \circ g)(x)$.

   (c) Find the domain and range of h.

   (d) (i) Show that $h^{-1}(x) = e^{\frac{1}{x-3}}$

       (ii) Solve the equation $h(x) = h^{-1}(x)$ for $x < 1$.

       (iii) Find the x-intercept of $h(x)$.

       (iv) Find an expression involving integrals for the area bounded by the x- and y-axes, and the graphs of h and h^{-1}.

       (v) Evaluate the integral in part (iv).

Answers:

(a) $y = 2 - \dfrac{1}{x-2}$

(b) $y = \dfrac{1}{\ln x} + 3$

(c) Domain is $x > 0$, $x \neq 1$. Range is $h(x) \neq 3$.

(d) (ii) $x = 0.653$

    (iii) $(0.717, 0)$

    (iv) Area $= \int_0^{0.653} e^{\frac{1}{x-3}} \, dx + \int_{0.653}^{0.717} \dfrac{1}{\ln x} + 3 \, dx$

    (v) 0.4706

Chapter 3: GEOMETRY AND TRIGONOMETRY

3.1 Solution of Triangles

Right-angled triangles: This first section is a reminder of how to deal with the sides and angles of a right-angled triangle. The following page deals with non right-angled triangles.

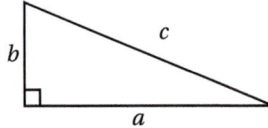

Pythagoras' Theorem: If you know two sides of a right-angled triangle, you can calculate the third using Pythagoras' Theorem. This states that the square of the hypotenuse (the longest side) equals the sum of the squares of the two shorter sides. As applied to the diagram, $c^2 = a^2 + b^2$. You must remember to subtract if you already have the hypotenuse (it's always opposite the right angle) and want to calculate one of the other sides. For example, $b^2 = c^2 - a^2$.

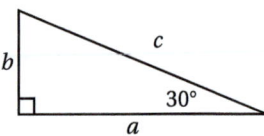

Trigonometry: There is no mystery to sin, cos and tan. They simply represent the ratios of pairs of sides for a triangle with given angles. For example, suppose the smallest angle in the triangle is 30°. Whatever the *size* of the triangle, b turns out to be half of c. The ratio of b to c is called the sine (sin for short), so $\sin 30° = 0.5$. The ratio of a to c is called the cosine (cos), and b to a is the tangent (tan). If you use the following procedure **in all cases** then every question can be worked out in the same way, and you should always get the right answer.

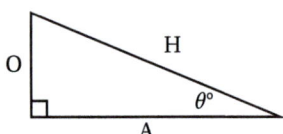

1. Label the three sides of the triangle with H (for hypotenuse, the side opposite the right angle), O (for opposite, the side opposite the angle you are dealing with) and A (for adjacent, the side next to the angle).
2. For the two sides you are dealing with, write down the word sin, cos or tan according to the mnemonic SOH/CAH/TOA.
3. Now write down the angle (which may be unknown) followed by an equals sign.
4. On the right hand side of the equals sign, you will write down a fraction (O over H, A over H or O over A) which will either involve two known sides, or one known and one unknown side.
5. You will now have an equation to solve. The three examples below show how to do this.

3. GEOMETRY AND TRIGONOMETRY

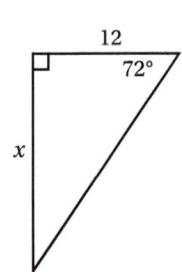

Example 1: Find x.

x is O, 12 is A, so we use tan.

Write down tan, then the angle, then =, then the fraction O/A.

To solve this equation, just multiply through by 12.

$\tan 72° = \dfrac{x}{12}$

$12 \times \tan 72° = x$

$x = 36.9°$

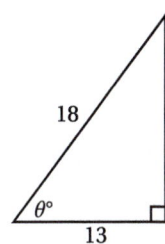

Example 2: Find s.

s is H, 7.5 is O, so we use sin.

Write down sin, then the angle, then =, then the fraction O/H.

The unknown is now on the bottom of the fraction, so we must cross-multiply to find s.

$\sin 35° = \dfrac{7.5}{s}$

$s = \dfrac{7.5}{\sin 35°}$

$s = 13.1$

Example 3: Find the angle $\theta°$.

13 is A, 18 is H, so we use cos.

Write down cos, then the angle, then =, then the fraction A/H.

Calculate the value of the fraction, then use the $\cos^{-1}$ function to find out the angle ($\cos^{-1}$ means "find the angle whose cosine is...)

$\cos \theta = \dfrac{13}{18}$

$\cos \theta = 0.7222$

$\theta = \cos^{-1} 0.7222$

$\theta = 43.8°$

Having worked out $\dfrac{13}{18}$ leave the answer on the display. Then work out the angle using $\cos^{-1}$ANS. This ensures full accuracy.

Sine and Cosine Rules: For triangles which are *not* right-angled we use the sine and cosine rules. The triangle shown has the conventional notation of small letters for the lengths of sides and capital letters for the angles opposite. To find lengths and angles, use:

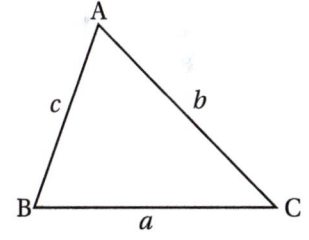

- The sine rule if 2 sides and 2 angles are involved, unless one of the angles is between the two sides
- The cosine rule if 3 sides and 1 angle are involved

SINE RULE	COSINE RULE
$\dfrac{a}{\sin A} = \dfrac{b}{\sin B} = \dfrac{c}{\sin C}$	$c^2 = a^2 + b^2 - 2ab \cos C$ *(for a side)* $\cos C = \dfrac{a^2 + b^2 - c^2}{2ab}$ *(for an angle)*

Don't be put off by the letters. Basically, the sine rule says the ratio of side/sine is the same for each pair of sides and angles. And in both versions of the cosine rule, ensure that the side c matches the angle C.

MATHEMATICS: ANALYSIS AND APPROACHES HL

In triangle ABC, angle B = 43°, AC = 6.8 cm and AB = 4.3 cm. Find the size of angle A giving your answer to the nearest degree.

$$\frac{\sin C}{4.3} = \frac{\sin 43}{6.8}$$

$$\sin C = \frac{4.3 \sin 43}{6.8} = 0.4313$$

$$\therefore C = 25.55°$$

So $A = 180 - (43 + 25.55) = 111.45$

$A = 111°$ to the nearest degree.

It's very useful to draw a quick sketch to see how to proceed.

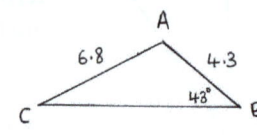

We know 2 sides and 1 angle, and we want another angle, so we use the sine rule (which I've inverted to make the calculation easier — always start by writing the thing you want to work out).

In this case we can only work out C from the sine rule, but then we can use the sum of the angles to find A. Note that you should always work to more figures than you need.

Example: A triangle has sides 4, $\sqrt{48}$ and 8. Calculate the size of the angle opposite the side with length $\sqrt{48}$.

Solution: This time we need the cosine rule in its second form, making sure that the side labelled c in the formula is opposite the required angle.

$$\cos C = \frac{4^2 + 8^2 - (\sqrt{48})^2}{2 \times 4 \times 8}$$. Check that this gives C = 60°.

Make life easy for yourself by remembering that you don't need a calculator to find the square of a square root! $(\sqrt{48})^2 = 48$.

Ambiguous case using the sine rule: Suppose we are given a triangle where AC = 8, BC = 5, and angle A = 30°. The diagram shows that with this information there are two possible triangles which can be drawn, and hence two possible values for angle B – this is known as the *ambiguous case*. Having found one answer, the other can be found by subtracting from 180°. In this case, B_1 is 126.9° and B_2 is 53.1°.

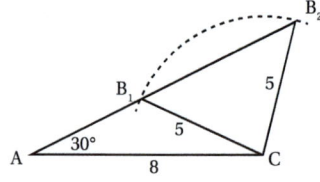

Area of a non-right angled triangle: If you know two sides of a triangle, and the size of the angle between the two sides, then the area of the triangle can be found using: Area = $\frac{1}{2} ab \sin C$.

But don't forget the alternative formula $A = \frac{1}{2} \times$ base $\times$ height which is useful if two sides are perpendicular.

Example: A triangle has sides 5, 7, and 8. Find the size of the smallest angle and the area of the triangle.

Solution: The smallest angle is opposite the shortest side. Using the cosine rule we get $\cos x = \frac{7^2 + 8^2 - 5^2}{2 \times 7 \times 8} = 0.786$.

This gives an angle of 38.2°. The sides either side of this angle are 7 and 8, so the area is $\frac{1}{2} \times 7 \times 8 \times \sin 38.2° = 17.3$

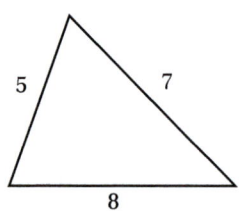

Bearings: One of the practical applications of non-right angled trigonometry is the calculation of distances and angles for moving ships and planes. Their direction of travel is based on compass directions, called *bearings*. A bearing is an angle measured around clockwise from North. Always draw in North lines on your diagrams before marking in bearings.

If a question involves bearings between places, check whether you are dealing with the bearing of A *from* B or the bearing from A *to* B, which is the other way round. Use arrows to show in which direction to take the bearing, and put the North line at the *start* of the arrow.

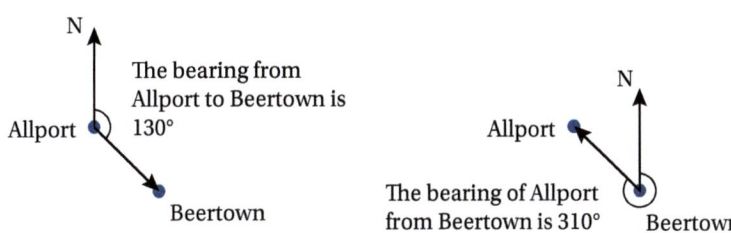

There is always a difference of 180° between bearings in opposite directions.

If you find calculations with bearings a bit confusing, I suggest you work through the following question with me. It's important that you work with a large, clear sketch. Also, at the risk of repeating myself, note that I have worked to 4SF in order to give accurate answers to 3SF.

> A ship sails from port P and travels due South to port Q. From port Q it sails on a bearing of 065° and travels for 45 km to a point R, which is due East of P.
>
> (a) (i) Draw and label clearly a diagram to show P, Q and R.
>
> (ii) Calculate the distance from port P to point R.
>
> *In questions like this the diagram is an important tool, so make it large. Angles do not have to be accurate or lengths drawn to scale, but make them look approximately right.*
>
> (i) [diagram: right triangle with P top-left, R top-right, Q bottom, 45km hypotenuse, 65° at Q]
>
> (ii) $\sin 65° = \dfrac{PR}{45} \Rightarrow PR = 45\sin 65° = 40.8$
>
> The distance from P to R is 40.8 km

A second ship also sails from port P for 45 km to a point S, but on a bearing of 330°.

(b) Complete your diagram in part (a) to show point S.

(c) Calculate the distance from R to S (shown with a grey dotted line) and the angle PRS.

(b)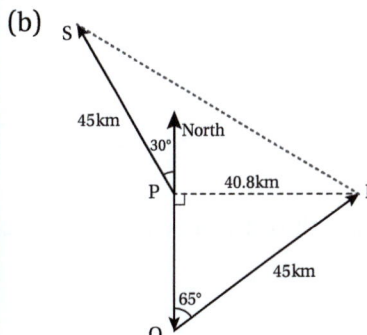

Rather than putting in 330°, the more useful 30° has been shown instead. The 40.8 has also been put in: always keep your diagrams up-to-date with new information.

To calculate RS, we use triangle PRS which is not right angled. We already know two sides and one angle (SPR = 30 + 90 = 120°), so we use the cosine rule:

MATHEMATICS: ANALYSIS AND APPROACHES HL

> (c) $RS^2 = 45^2 + 40.783^2 - 2 \times 45 \times 40.783 \times \cos 120°$
>
> $RS = \sqrt{5523.5} = 74.3 \text{ km}$
>
> *(Check: RS < RP + PS. Looks OK)*
>
> Now we need to calculate angle PRS. We know one angle and two sides so we use the sine rule.
>
> $\dfrac{\sin PRS}{45} = \dfrac{\sin 120}{74.321} \Rightarrow \sin PRS = \dfrac{45 \sin 120}{74.321} = 0.5244$
>
> So angle PRS = $\sin^{-1}(0.5244) = 31.6°$

> (d) What is the bearing of S from R?
>
>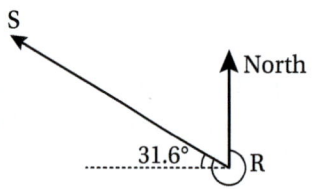
>
> The diagram shows the arrow representing S from R, and a new North line inserted. The required bearing has also been put in. How big is this angle? From North round to West is 270°, and then we need another 31.6.
>
> (d) The bearing of S from R is 270 + 31.6 = 301.6°

Answers

1. $b = 5.32$ cm, $c = 7.20$ cm
2. $p = 4.31$ cm, Q = 79.9°
3. 80.3 m, 2.87°
4. 10.9 km, 059.4°
5. 14.5 cm
6. A = 36.9°, B = 90°, C = 53.1°
7. 9.64 cm
8. 7.93

Solution of Triangles: Practice Exercise

1. Triangle ABC has A = 66°, B = 44°, a = 7cm. Find b and c.

2. Triangle PQR has P = 45°, r = 5cm and q = 6cm. Find p and Q.

3. a) An observer on a ship which is 2.3km from the coast measures the angle of elevation of a cliff as 2°. Find the height of the cliff in metres.

   b) A 35m tower stands on the top of the cliff. Find the angle of depression for an observer looking at the ship from the top of the tower.

   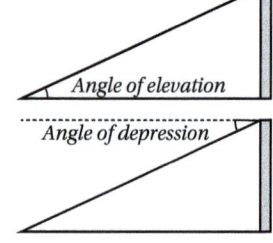

4. A ship S is 7km away from a lighthouse L on a bearing of 080° and a ship T is 5km away from L on a bearing of 210°. Find the distance and bearing of S from T.

5. A rhombus has sides of length 8cm and angles of 50° and 130°. Find the length of the longer diagonal of the rhombus.

6. A triangle ABC has area 24 cm² and sides a = 6cm, b = 10cm. Find all the angles in the triangle.

7. Calculate WX, given YZ = 15 cm.

   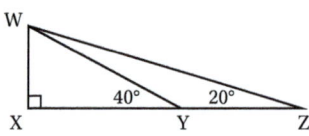

8. In the triangle ABC, angle A = 30°, a = 4 and c = 6. Find the difference in area between the two possible triangles ABC.

3. GEOMETRY AND TRIGONOMETRY

3.2 3-D Geometry

You have to be able to combine the rules of 3-D coordinates with the trigonometry of both right angled and non-right angled triangles in order to find the sides and angles of cuboids, prisms and pyramids.

Cuboid: A cuboid is the 3-D equivalent of a rectangle. It has 12 edges, 4 each in three different dimensions (length, width and height). Commonly asked questions involve the lengths of diagonals (both of sides and also from one corner to the opposite corner) and angles between various lines. The points used will be the vertices (corners) and the midpoints of sides.

Volume of a cuboid: $V = $ length × width × height

Pyramid: You only have to concern yourself with a "right" pyramid – ie where the apex is directly above the centre of the base, which is itself a square. Pyramid questions almost invariably use the midpoints of sides, and it should be noted that a line drawn from the midpoint of one of the base edges to the apex is at a right angle to the base edge.

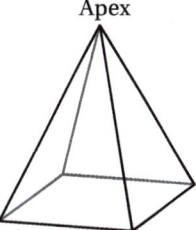

Volume of a pyramid: $V = \frac{1}{3} \times$ base area × height

Prism: A prism is any 3-D shape with the same cross-section throughout its length. Very often this cross-section is a triangle, but it does not have to be.

Volume of a prism: $V = $ area of cross section × length

Angle between a line and a plane: A plane is a flat surface, so each of the faces of the 3-D shapes illustrated is a plane.

The angle between a line and a plane is the angle between the line and its *projection* on the plane: think of the projection as part of a "shadow line" on the plane.

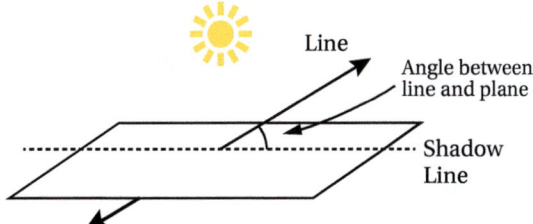

Calculating lengths and angles in 3 dimensions: In every case, you convert the question into an appropriate 2-D question, usually by identifying a right-angled triangle containing the length/angle you have to work out, drawing it as it really looks, then using trigonometry and Pythagoras as usual.

For example, the base edges of a pyramid are 10 cm and the slant edges are 12 cm. M is the midpoint of side PQ and X is the centre point of the base. Find length AM and angle AMX to the nearest degree.

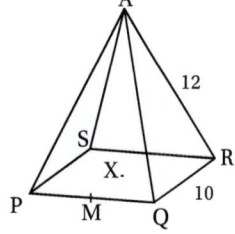

MATHEMATICS: ANALYSIS AND APPROACHES HL

We can find AM using triangle AMQ.

MQ = 5 (half of the base length)

So $AM^2 = 12^2 - 5^2 \Rightarrow AM = \sqrt{119}$.

> No need to calculate $\sqrt{119}$ since we are using it in the next stage.

Now we can draw triangle AMX because we know MX = 5 and $AM = \sqrt{119}$.

We can see from the diagram that $\cos M = \dfrac{5}{\sqrt{119}}$ giving M = 62.7°, or 63° to the nearest degree.

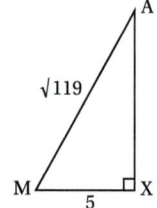

A prism ABCDEF is formed with an isosceles triangle cross section and a rectangular base. AE = 6.4 cm, AB = 5 cm, BC = 11 cm. M is the midpoint of AB.

(a) Find the length of EM.

(b) Hence find

    (i) The area of triangle AEB

    (ii) The volume of the prism

(c) Find the length of MC, and hence the angle EC makes with base ABCD.

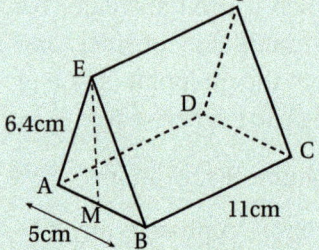

a) $EM^2 + 2.5^2 = 6.4^2$

$EM^2 = 34.71$

$EM = 5.89$ cm

b) (i) Area = $\tfrac{1}{2}$ × base × height = $\tfrac{1}{2}$ × 5 × 5.89 = 14.7 cm²

   (ii) Volume = 14.7 × 11 = 162 cm³

c) $MC^2 = 2.5^2 + 11^2$

$MC = 11.3$ cm

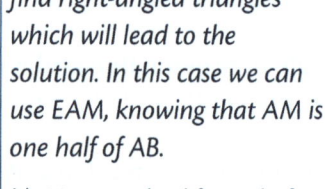

$\text{Tan}(ECM) = \dfrac{5.89}{11.3}$

Angle EC makes with base is 27.6°

a) Wherever possible, try to find right-angled triangles which will lead to the solution. In this case we can use EAM, knowing that AM is one half of AB.

b) Use standard formula for the area of a triangle and the volume of a prism.

c) Once again we find in EMC a right-angled triangle which will give us MC. It also gives us the angle EC makes with the base.

As before, I have given answers to 3SF, but worked to 4SF.

3. GEOMETRY AND TRIGONOMETRY

3.3 Cylinder, Sphere and Cone

Curved Surface Area: Whilst the concept of the *volume* of shapes with curved edges and faces may not be too difficult to appreciate, the *area* of such curved faces may be problematic. The way to think of such faces is to imagine them to be made from separate pieces of paper; the area of a curved surface is the same as the area of the paper it is made from when it is flattened out. For example, unroll the curved surface of a cylinder and you get a rectangle whose height is the height of the cylinder and whose length is the circumference of the cylinder.

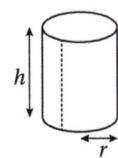

Cut along the dotted line and open up to get...

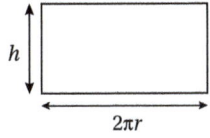

You need to be able to use the relevant area and volume formulae for cylinders, cones and spheres. You will find them all in your list of formulae. They are:

Cylinder: Curved surface area = $2\pi r h$

Volume = $\pi r^2 h$

Cone: Curved surface area = $\pi r l$ (l is the slant height)

Volume = $\frac{1}{3}\pi r^2 h$

Sphere: Surface area = $4\pi r^2$

Volume = $\frac{4}{3}\pi r^3$

Read each question very carefully to see *exactly* what you are being asked to find. For example, a cylinder may be completely closed in which case the total surface area is the curved surface area plus the areas of the two ends, which are both circles. Or it may be open at one end, so just add one circle.

The volume of a hemisphere is half that of a sphere, but its total surface area will be half the curved surface of a sphere plus the circle which forms its base. Its formula will be $\frac{1}{2} \times 4\pi r^2 + \pi r^2 = 3\pi r^2$.

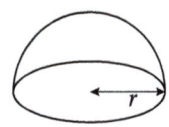

Example: Both a cone and a hemisphere have base diameter of 18 cm. If the height of the cone is 10 cm, show that the ratio of the volume of the cone to that of the hemisphere is 5:9.

Solution: The volume of the cone is $\frac{1}{3}\pi r^2 h = \frac{1}{3}\pi \times 9^2 \times 10 = 270\pi$. The volume of the hemisphere is $\frac{1}{2} \times \frac{4}{3}\pi r^3 = \frac{2}{3}\pi \times 9^3 = 486\pi$. We may as well leave π in the answers since we are finding the ratio.

So, volume of cone:volume of hemisphere = $270\pi : 486\pi$ = 5:9

> There's an old trap in this question - all the formulae use the radius, but we have been given the diameter.

Try the following example for yourself.

Example: In the South of England, special brick constructions called "Oast Houses" were built to dry the hops used in making beer. The diagram models an Oast House as a cone on top of a cylinder. Both have base diameter 8 m; the cylinder has a height of 6 m, and the building is 17.4 m high overall.

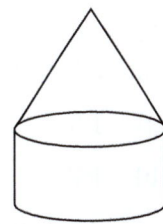

Find the total surface area of the exposed faces of the Oast House, and the total volume.

MATHEMATICS: ANALYSIS AND APPROACHES HL

You will need the slant height of the cone to find the curved surface area – use Pythagoras' Theorem. Note too that the base of the cone will not form part of the total surface area since it is not an external surface.

> Full working can be found on the website

Answers: Area = 303 m², Volume = 493 m³.

A cylindrical tube of length 40 cm and radius 3.2 cm contains 6 tennis balls each with a radius of 3.14 cm.

(a) Giving answers to 3SF, find

 (i) The volume of the tube

 (ii) The volume of one tennis ball

(b) Hence find the volume of empty space in the tube.

(a) (i) $V = \pi r^2 l = \pi \times 3.2^2 \times 40 = 1286.7...$

Volume of tube = 1290 cm³

 (ii) $V = \frac{4}{3}\pi r^3 = \frac{4}{3}\pi \times 3.14^3 = 129.68...$

Volume of ball = 130 cm³

(b) Volume of 6 tennis balls = 129.68 × 6

= 778.1 cm³

Empty space = 1286.7 – 778.1 = 508.6...

= 509 cm³ to 3SF

If you're going to use a formula, write it down. This shows the examiner what you are using, and helps you to substitute values correctly.

This question is yet another good example of the importance of working to a higher accuracy than the required answer. If the first two rounded answers were used to obtain the final answer, the result would have been

1290 – 6 × 130 = 510.

Geometry: Practice Exercise

> **Answers**
> 1. 5.44 km, 272.2°
> 2. 10 cm, 13 cm, 67.4°, 12.4 cm, 76.0°, 5 cm, 40.6°
> 3. Volume of cone = 25.1 cm³
> Volume of each sphere = 0.524 cm³
> 47 spheres, with a bit left over!

1. Three boats P, Q and R are at anchor in a bay. The bearing of P from R is 046°, and of Q from P is 125°. The distance of R from P is 3 km, and of P from Q is 4 km.

 a) Draw a clear diagram showing all the information.

 b) Calculate:

 i) the distance of R from Q;

 ii) the bearing of R from Q.

2. The diagram shows a right pyramid with:

 PQ = 8 cm, QR = 6 cm, VW = 12 cm.

 W is at the centre of the base, M and N are the midpoints of PQ and QR. Calculate:

 a) PR

 b) PV

 c) Angle VPW

 d) VM

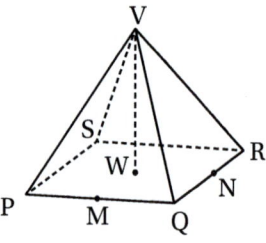

74

e) The angle VM makes with the base

f) MN

g) Angle MSN *(Draw triangle MSN first)*

3. A solid cone of height 6 cm and radius 2 cm is melted down to be made into small spheres. If each sphere has a radius of 0.5 cm, how many can be made?

3.4 Circular Functions

Radians: Radians are an alternative to degrees when measuring the size of angles. Although it is easier to *think* in degrees, radians are often used with trigonometric functions and *must* be used when differentiating or integrating them.

The conversion is π radians = 180°. (An angle is assumed to be in radians unless the degrees symbol is given.)

It is worth memorising some key angles in radians (see table in the notes box). π appears in many angles when expressed in radians (because of the conversion) but it does not have to. For example, 45° = 0.785 radians, but this is not an *exact* conversion, unlike $\frac{\pi}{4}$.

There are two circle formulae which are used when a sector angle is expressed in radians. If the angle is q and the radius of the circle is r:

- Arc length of sector = $r\theta$
- Area of sector = $\frac{1}{2}r^2\theta$

📱 *Before* making a start on any trigonometric question, check your calculator is in the right mode.

$30° = \pi/6$
$45° = \pi/4$
$60° = \pi/3$
$90° = \pi/2$
$120° = 2\pi/3$
$180° = \pi$
$270° = 3\pi/2$
$360° = 2\pi$

> The diagram shows two concentric circles with radii 1 and 4. The area S is enclosed by arcs AB and CD, and lines AD and BC.
>
> Given that angle AOB = $\frac{\pi}{3}$, find:
>
> (a) The area of S
> (b) The perimeter of S
>
>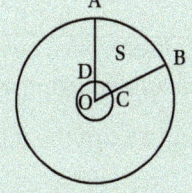
>
> simplifying your answers and giving them in an exact form.

(a) Area of sector AOB = $\frac{1}{2}r^2\theta = \frac{1}{2} \times 4^2 \times \frac{\pi}{3}$

$= \frac{16\pi}{6} = \frac{8\pi}{3}$

Area of sector DOC = $\frac{1}{2} \times 1^2 \times \frac{\pi}{3}$

$= \frac{\pi}{6}$

Area of S = Area AOB − Area DOC

$= \frac{8\pi}{3} - \frac{\pi}{6} = \frac{15\pi}{6}$

(b) Perimeter of S = AD + BC + arc AB + arc DC

$= 3 + 3 + \frac{4\pi}{3} + \frac{\pi}{3}$

$= 6 + \frac{5\pi}{3}$

Often an odd shaped area is found by subtracting two friendly shaped areas!

Notice how I have used words to explain to the examiner what each calculation is about, rather than just an amorphous mass of calculations. And, in case you've done your sums wrong, it's important to state that you're subtracting the two areas, and what you've added to get the perimeter. There are method marks for these, even if all the numbers are wrong!

Trigonometric functions: The diagram shows a circle with radius 1 (a *unit circle*). A line is drawn from the centre to a point on the circumference, and this forms angle θ with the x-axis. The x-coordinate of the point is defined as the cosine of the angle ($\cos\theta$) and the y-coordinate as the sine ($\sin\theta$).

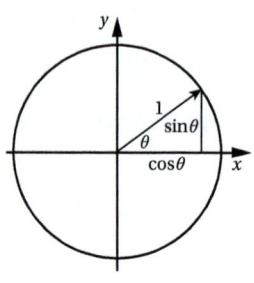

These definitions will apply as the line rotates full circle, giving the sin and cos for all angles from 0° to 360°.

> These graphs can, of course, be extended to show the sin and cos for *all* angles.

When these are plotted as graphs, we get the following:

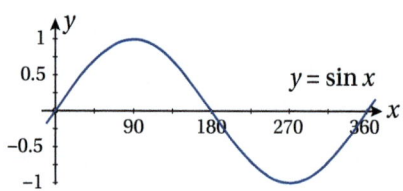 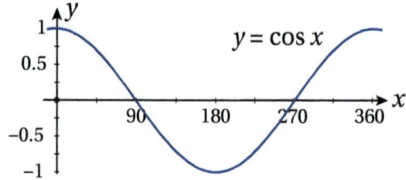

Points to note:

- The range of both functions is $-1 \leq f(x) \leq 1$
- $\sin x > 0$ for angles between 0° and 180°
- $\cos x > 0$ for angles between 0° and 90°, also between 270° and 360°
- Both functions have a *period* (ie repeat themselves) every 360°.

The graph of $\tan x$ is still *periodic* but with a period of 180° rather than 360°. It also has vertical asymptotes at 90°, 270° and so on.

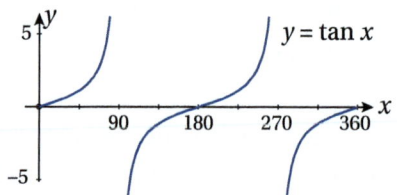

Exact values of sin, cos and tan: Although the trigonometric ratios for most angles have to be calculated using a GDC, some key angles have easily remembered values.

You should learn the following table:

> A neat way to remember the exact values for sin (ie the third row of the table) is that they form the series $\frac{\sqrt{0}}{2}, \frac{\sqrt{1}}{2}, \frac{\sqrt{2}}{2}, \frac{\sqrt{3}}{2}, \frac{\sqrt{4}}{2}$.

Angle in degrees	0	30	45	60	90
Angle in radians	0	$\frac{\pi}{6}$	$\frac{\pi}{4}$	$\frac{\pi}{3}$	$\frac{\pi}{2}$
sin	0	$\frac{1}{2}$	$\frac{\sqrt{2}}{2}$	$\frac{\sqrt{3}}{2}$	1
cos	1	$\frac{\sqrt{3}}{2}$	$\frac{\sqrt{2}}{2}$	$\frac{1}{2}$	0
tan	0	$\frac{1}{\sqrt{3}}$	1	$\sqrt{3}$	∞

Also note that $\frac{\sqrt{2}}{2} = \frac{1}{\sqrt{2}}$; choose whichever form is more convenient in a particular question.

You can now use this table, combined with the symmetries of the graphs, to calculate trigonometric ratios for angles greater than 90°. So, if I needed to find the value of $\sin 210°$ I would note from the graph that it is the same as $\sin 30°$ but with a negative sign. Thus, $\sin 210° = -\frac{1}{2}$.

Try the following, and check your answers on your GDC:

$$\cos 135°; \quad \sin\frac{3\pi}{4}; \quad \sin 180°; \quad \tan\frac{4\pi}{3}; \quad \tan 225°; \quad \cos\frac{11\pi}{6}.$$

Instead of using graph symmetries, you may have been shown how to use unit circles to deal with angles greater than 90°. See the website for a full explanation.

Simple trigonometric equations: What is the value of θ given $\sin\theta = 0.5$, $0° \leq \theta \leq 360°$? We want to know what angle has a sin which is 0.5, and this is 30°; then, using the symmetry of the sin graph, we can see that 150° is also a solution. (If the domain is in radians, you can either work in degrees and convert at the end, or set your calculator to radians: this gives $\theta = \frac{\pi}{6}$ or $\frac{5\pi}{6}$. In paper 2, a simple way to solve the equation is to find the intersections of $y = \sin x$ and $y = 0.5$ within the given domain.

Finding cos and tan from sin: A simple trick is to draw a right-angled triangle. If $\sin\theta = \frac{3}{5}$, what is $\cos\theta$? Having put 3 as the "opposite" and 5 as the hypoteneuse, the remaining side must be 4 (by Pythagoras). So $\cos\theta = \frac{4}{5}$ and $\tan\theta = \frac{3}{4}$. If θ was obtuse (that is, between 90° and 180°), $\cos\theta$ would be $-\frac{4}{5}$.

$\tan\theta$ is calculated as $\frac{\sin\theta}{\cos\theta}$ and has period π.

The reciprocal trigonometric functions: Each of sin, cos and tan has a reciprocal function. The periods of these functions are the same as the functions they are linked to:

- **sec** θ is the reciprocal of $\cos\theta$ ('sec' is short for *secant*)
- **cosec** θ is the reciprocal of $\sin\theta$ ('cosec' is short for *cosecant*)
- **cot** θ is the reciprocal of $\tan\theta$ ('cot' is short for *cotangent*)

The third letter of each of the reciprocal functions tells you to which function it is linked.

*se**c** with **c**os*
*cose**c** with **s**in*
*co**t**an with **t**an*

Example: Simplify $\dfrac{\cos^2\theta \sec\theta}{\sin\theta}$

Solution: Often the best way to deal with this sort of question is to reduce everything to sin and cos.

$$\frac{\cos^2\theta \sec\theta}{\sin\theta} = \frac{\cos^2\theta \times \frac{1}{\cos\theta}}{\sin\theta} = \frac{\cos\theta}{\sin\theta} = \cot\theta$$

Without using a calculator, find the exact values of:

$$\sec 45°, \cot\frac{\pi}{4}, \csc\frac{\pi}{6}, \sec 120°, \csc\frac{3\pi}{4}, \csc(-120°)$$

Answers:
$\sqrt{2}, 1, 2, -2, \sqrt{2}, -\frac{2}{\sqrt{3}}$

Without using a calculator, solve these equations for $0 \leq x \leq 2\pi$:

$$\sec\theta = 2, \quad \cot 2\theta = \sqrt{3}, \quad 3\cot\theta = \tan\theta$$

Answers:
$\frac{\pi}{3}, \frac{5\pi}{3}; \frac{\pi}{12}, \frac{7\pi}{12}, \frac{13\pi}{12}, \frac{19\pi}{12};$
$\frac{\pi}{3}, \frac{2\pi}{3}, \frac{4\pi}{3}, \frac{5\pi}{3}$

Did you get all four answers to the last part? Don't forget to take the negative square root as well as the positive.

See page 80 for hints on solving equations with 2θ.

3.5 Trigonometric Identities

You are provided with a whole range of trigonometric identities which, although available in the exam, are best learnt. Otherwise you may not recognise when to use a particular identity. What do we use them for?

- Simplifying trigonometric expressions
- Solving trigonometric equations
- Helping to integrate and differentiate awkward functions
- Working out *more* trigonometric identities!

MATHEMATICS: ANALYSIS AND APPROACHES HL

- The identities divide comfortably into groups.

The Pythagorean identities: The first two of these tend to appear more often than the third.

$$\sin^2\theta + \cos^2\theta \equiv 1$$
$$1 + \tan^2\theta \equiv \sec^2\theta$$
$$1 + \cot^2\theta \equiv \csc^2\theta$$

> Except for linear functions, you must remember that $f(a + b) \neq f(a) + f(b)$ although such expressions can often be expanded in other ways. Which of these can be written in other ways?
> $\sin(a + b)$
> $(a + b)^2$
> $\sqrt{(a + b)}$
> $\ln(a + b)$
> e^{a+b}
> (Answers: 1, 2 and 5)

The addition (compound) formulae: You must never simplify an expression such as $\sin(30° + x)$ as $\sin 30° + \sin x$. The six addition (compound) formulae which enable you to remove the brackets are:

$$\sin(A \pm B) \equiv \sin A \cos B \pm \cos A \sin B$$
$$\cos(A \pm B) \equiv \cos A \cos B \mp \sin A \sin B$$
$$\tan(A \pm B) \equiv \frac{\tan A \pm \tan B}{1 \mp \tan A \tan B}$$

The double angle formulae: The equation $2\sin x = \sin 2x$ cannot be solved directly because it is not possible to work with the x and the $2x$ at the same time. The double angle formulae help us to simplify such expressions.

Note that there is only one formula for $\sin 2\theta$, three for $\cos 2\theta$ and one for $\tan 2\theta$:

$$\sin 2\theta \equiv 2 \sin\theta \cos\theta$$
$$\cos 2\theta \equiv \cos^2\theta - \sin^2\theta$$
$$= 1 - 2\sin^2\theta$$
$$= 2\cos^2\theta - 1$$
$$\tan 2\theta \equiv \frac{2\tan\theta}{1 - \tan^2\theta}$$

Symmetry relationships: In general, transforming the graph of $f(x)$ into the graph of $f(a - x)$ is achieved by a reflection in the line $x = \frac{1}{2}a$. Consider the reflectional symmetry of the graphs of $\sin x$, $\cos x$ and $\tan x$ (see page 76):

> These relationships can also be derived from consideration of the unit circle.

- $\sin x$ is a reflection of itself around $x = \frac{\pi}{2}$ and so $\sin(\pi - \theta) \equiv \sin\theta$
- $\cos x$ reflected in $x = \frac{\pi}{2}$ gives $-\cos x$ and so $\cos(\pi - \theta) \equiv -\cos\theta$
- $\tan x$ reflected in $x = \frac{\pi}{2}$ gives $-\tan x$ and so $\tan(\pi - \theta) \equiv -\tan\theta$

How are the formulae used: Much of the algebra you learn enables you to simplify expressions, or rewrite them in different ways, with the aim of solving equations, differentiating or integrating. The identities listed above extend this idea into trigonometry, and the examples in the next section show how they can be used to help solve trigonometric equations.

> 📱 This question is an ideal case where you should check your answer on the GDC (if it's a Paper 2 question). Find x using arcsin$x$, then find sin$2x$. Check this decimal answer is the same as the exact answer.

Example: Given that $\sin x = \frac{5}{8}$, find an exact value for $\sin 2x$ if $0 \leq x \leq \frac{\pi}{2}$.

Solution: The double angle formula tells us that $\sin 2x = 2\sin x \cos x$, and we can use the little trick on page 77 to find $\cos x$; or we can use the Pythagorean identity $\cos^2 x \equiv 1 - \sin^2 x$.

$$\cos^2 x = 1 - \frac{25}{64}$$
$$= \frac{39}{64}$$
$$\cos x = \frac{\sqrt{39}}{8}$$
$$\text{So } \sin 2x = 2 \times \frac{5}{8} \times \frac{\sqrt{39}}{8}$$
$$= \frac{5\sqrt{39}}{32}$$

If $\frac{\pi}{2} \le x \le \pi$, we would have used $\cos x = -\frac{\sqrt{39}}{8}$.

3.6 Solving Trigonometric Equations

The basics: Because all trigonometric functions are periodic there will always be an infinite number of solutions to any trig equation; that is why all questions will restrict the range of solutions – make sure you take note of the given range. In all the examples that follow, the solutions are such that $0° \le x \le 360°$. If the range is given in radians, your answers must be in radians.

REMEMBER:
Always check the mode setting of your calculator *before* solving trigonometric equations.

An example of the simplest equations is given on page 77. Suppose you are asked to solve $2\sin 2x + 2 = 1$. There will be 4 solutions (because of the $2x$), and the way to find them is:

Reduce the equation so that the LHS just has the sin function	$\sin 2x = -0.5$
Calculate $\arcsin(-0.5)$	210° and 330°
Find the next two angles by adding 360°	570° and 690°
These will be values of $2x$, so the four solutions for x are	105°, 165°, 285°, 345°

Using the identities: The following examples show how the identities can be used in a variety of situations. In each case, we are aiming to get an equation with just one trig function in it which can then be solved as above. The last part is left for you to complete in each case!

You can of course use your GDC in Paper 2; but the question may be put in such a way that you must use the identities. In example 1 below, an exam question may say: "Show that this equation reduces to $\tan \theta = \frac{3}{4 - 3\sqrt{3}}$."

Example 1 *Using an addition formula. Easily recognisable because of the brackets.*

Find the value of $\tan \theta$ if $2 \sin \theta = 3 \cos(\theta - 60°)$

$$2 \sin \theta = 3 \cos \theta \cos 60° + 3 \sin \theta \sin 60°$$
$$2 \sin \theta = \frac{3}{2} \cos \theta + \frac{3\sqrt{3}}{2} \sin \theta$$
$$4 \sin \theta = 3 \cos \theta + 3\sqrt{3} \sin \theta$$
$$(4 - 3\sqrt{3}) \sin \theta = 3 \cos \theta$$
$$\tan \theta = \frac{3}{4 - 3\sqrt{3}}$$

Note how we can get a tan function from here. We couldn't if there was an extra numerical term.

MATHEMATICS: ANALYSIS AND APPROACHES HL

Example 2 *Using a double angle formula. An equation cannot be solved if it has functions of both θ and 2θ in it unless a double angle formula is used.*

$$\sin 2\theta = \sqrt{2}\sin\theta$$
$$2\sin\theta\cos\theta = \sqrt{2}\sin\theta$$
$$2\sin\theta\cos\theta - \sqrt{2}\sin\theta = 0$$
$$\sin\theta(2\cos\theta - \sqrt{2}) = 0$$
$$\sin\theta = 0 \text{ or } \cos\theta = \frac{\sqrt{2}}{2}$$

← Don't divide by $\sin\theta$ here; you will lose some solutions. Always factorise.

Solutions: $\theta = 0°, 45°, 180°, 315°, 360°$

Example 3 *Using the $\tan 2\theta$ formula – often a case of 'try it and see.'*

$$\tan\theta \tan 2\theta = 1$$
$$\tan\theta \times \frac{2\tan\theta}{1 - \tan^2\theta} = 1$$
$$2\tan^2\theta = 1 - \tan^2\theta$$
$$\tan^2\theta = \frac{1}{3}$$
$$\tan\theta = \pm\sqrt{\frac{1}{3}}$$

← Multiply both sides by $1 - \tan^2\theta$

← If you forget the negative square root you will lose two solutions.

Solutions: $\theta = 30°, 150°, 210°, 330°$

Example 4 *Using a $\cos 2\theta$ formula – often this leads to a quadratic equation.*

$$2 + \cos 2\theta = \sin\theta$$
$$2 + (1 - 2\sin^2\theta) = \sin\theta$$
$$2\sin^2\theta + \sin\theta - 3 = 0$$
$$(2\sin\theta + 3)(\sin\theta - 1) = 0$$
$$\sin\theta = -\frac{3}{2} \text{ or } \sin\theta = 1$$

← This is the most suitable version of the $\cos 2\theta$ formula because of the sin already in the equation.

← It may help to think of this as $2s^2 + s - 3 = 0$

Only possible solution: $\theta = 90°$

If you can, and if you have time, always substitute at least one of your answers into the original equation. If you find it doesn't work, first check that you have copied the question correctly, then check through the working carefully, paying particular attention to minus signs.

Trigonometric Formulae: Practice Exercise

If you get stuck with any of these, go the website for some hints to lead you to the solutions.

1. Solve $\cos 2\theta = 2\sin^2\theta$ for $0 \le \theta \le 2\pi$.
2. Solve $3\sin\theta = \tan\theta, 0 \le \theta \le \frac{\pi}{2}$.
3. Solve $\sec^2\theta = 8\cos\theta$ where θ is in the range $[-180°, 180°]$.
4. $\cos 2x + 7\sin x = 4$. ▶ *See video for how to derive a general solution*
5. $4 - \sin\theta = 4\cos^2\theta, 0° \le \theta \le 180°$.
6. Prove the following identities:
   (a) $\sin\theta\tan\theta + \cos\theta \equiv \sec\theta$
   (b) $\tan(A + B) \equiv \frac{\tan A + \tan B}{1 - \tan A \tan B}$ using the formulae for $\sin(A + B)$ and $\cos(A + B)$.
   (c) $\sqrt{\left(\frac{1 - \cos\theta}{1 + \cos\theta}\right)} \equiv \text{cosec}\theta - \cot\theta$ by first multiplying both lines of the fraction in the square root by $1 - \cos\theta$.

Answers
1. $\frac{\pi}{6}, \frac{5\pi}{6}$
2. $0, 1.23, \pi$
3. $-60°, 60°$
4. $\frac{\pi}{6} + 2\pi k, \frac{5\pi}{6} + 2\pi k, k \in \mathbb{Z}$
5. $0°, 14.5°, 165.5°, 180°$

3. GEOMETRY AND TRIGONOMETRY

> Show that $\arctan\left(\frac{1}{2}\right) + \arctan\left(\frac{1}{3}\right) = \frac{\pi}{4}$.

$\tan\left(\arctan\left(\frac{1}{2}\right) + \arctan\left(\frac{1}{3}\right)\right)$

$= \dfrac{\tan\left(\arctan\left(\frac{1}{2}\right)\right) + \tan\left(\arctan\left(\frac{1}{3}\right)\right)}{1 - \tan\left(\arctan\left(\frac{1}{2}\right)\right)\tan\left(\arctan\left(\frac{1}{3}\right)\right)}$

$= \dfrac{\frac{1}{2} + \frac{1}{3}}{1 - \frac{1}{2} \times \frac{1}{3}}$

$= \dfrac{5/6}{5/6} = 1$

Also $\tan\left(\frac{\pi}{4}\right) = 1$ so $\tan(\text{LHS}) = \tan(\text{RHS})$

$\therefore \text{LHS} = \text{RHS}$

The only way to deal with an arctan in an equation is to take its tan (like squaring a square root). But we can't take the tan of each term separately, so we try taking the tan of the LHS as a whole. This leads to an expression of the form tan(A + B), and we can then see what happens if we use the compound angle formula.

Another "try it and see what happens" question.

3.7 Graphing Periodic Functions

The connection between functions and transformations is covered on page 43. The functions $a\sin b(x + c) + d$ and $a\cos b(x + c) + d$ are a little special because each of the constants a, b, c and d have specific meanings which relate to "real-life" functions. The examples below use degrees for greater understanding, but most of the exam questions will use radians.

Let's consider how we can graph the function $f(x) = 2\sin 3(x + 20°) + 1$ as a transformation of the graph of $y = \sin x$. In particular, note that each of the constants has a transformation effect which is independent of the others. Starting inside the bracket:

- The 20 translates the graph –20° horizontally.
- The 3 makes the graph cycle 3 times faster than $y = \sin x$, ie the period is 120° rather than 360°. The *frequency* is therefore multiplied by 3.
- The 2 multiplies the *amplitude* by 2, ie the range of y is –2 to 2 rather than –1 to 1.
- The 1 translates the graph vertically, so the range of y become –1 to 3.

The period is calculated as $\frac{360}{b}$ or $\frac{2\pi}{b}$.

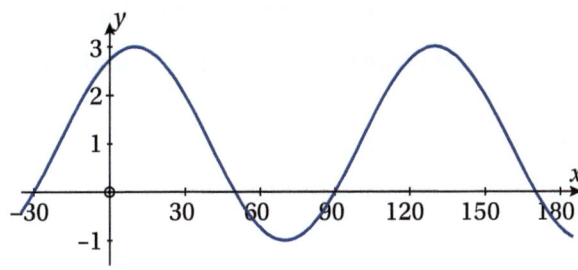

Sketching a graph such as this can also help us to see where solutions to equations of the form $a\sin(b(x + c)) + d = 0$ are going to occur. Let's see a practical situation – questions on periodic functions often involve the heights of tides which, because of the circular motion of the moon, are based on sin functions. In this question, however, we are using a periodic function to model a wave:

MATHEMATICS: ANALYSIS AND APPROACHES HL

A wave of water passing a fixed point is modelled by the function $h = 0.5\cos(0.25\pi t) + 1.8$ where h is the height of the water (in metres) above the point and t is the time (in seconds).

i) What is the period of the wave?

The period of the wave is $\dfrac{2\pi}{0.25\pi} = 8$s.

ii) Draw a sketch of the function for $0 \leq t \leq 16$.

We start with $\cos x$. The period is 8. The centre of the oscillation is 1.8 and amplitude 0.5.

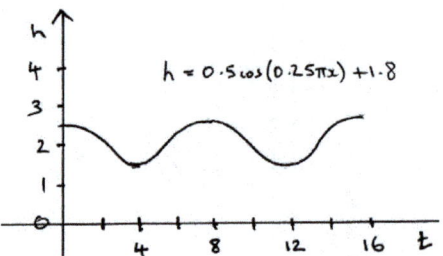

iii) Find the times for $0 \leq t \leq 16$ for which the wave height is 2.15 m.

Consider the horizontal line $h = 2.15$. It will intersect the graph 4 times, so we are looking for 4 solutions.

$$0.5\cos(0.25\pi t) + 1.8 = 2.15$$
$$\cos(0.25\pi t) = 0.7$$
$$0.25\pi t = 0.795, 5.488, 7.078, 11.77$$

So $t = 1.01$ s, 6.99 s, 9.01 s, 15.0 s

The values in the third line of working start with $\arccos 0.7 = 0.795$, then $2\pi - 0.79$. The next two solutions are then found by adding 2π to the first two.

3.8 Basics of Vectors

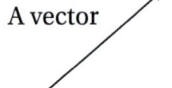

A vector

Same length, different direction – so a different vector

Notation: Think of a vector as representing a movement, or displacement, in a plane. This can be represented by an arrow. The vector can be defined in several ways:

- Using a single small letter. Bold type in printed text, line underneath or arrow on top in handwriting.
- Using the named points at either end, arrow on top.
- Using a "column vector" to show the displacement in the x, y and z (if applicable) directions.
- In the form $a\boldsymbol{i} + b\boldsymbol{j}$ where $\boldsymbol{i}$ and $\boldsymbol{j}$ are unit vectors in the x and y directions (this is equivalent to the column vector form but less easy to use).

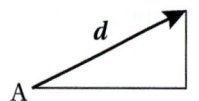

This vector could be written as:

$\overrightarrow{AB}$, $\boldsymbol{d}$, $\begin{pmatrix} 6 \\ 3 \end{pmatrix}$, $6\boldsymbol{i} + 3\boldsymbol{j}$ (as examples)

The column vector form is useful because we can work out the length and direction of the vector using Pythagoras and arctan (or using appropriate calculator functions).

Position and displacement vectors: If a vector is used to define the position of a point then it is known as a *position vector*. It will always start at the origin. The components of

3. GEOMETRY AND TRIGONOMETRY

the column vector will always be the same as the coordinates of the point. *Displacement* vectors differ from position vectors in that they have no specific position – they just represent a *change* in position.

Operating with vectors: If you move along a vector *a* then along a vector *b*, the single displacement which takes you to the end position is defined as vector *a* + *b*. The length of *a* + *b* is **not** the length of *a* plus the length of *b*; it is shorter. However, if *a* and *b* are written as column vectors, then adding them will give vector *a* + *b*. For example: $\binom{6}{-2} + \binom{-4}{5} = \binom{2}{3}$.

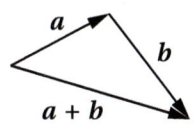

A vector can be *multiplied* by a number. For example, 2*a* has the same direction as *a* but is twice as long. Using column vectors, eg: $2\binom{3}{5} = \binom{6}{10}$. And a *minus* sign reverses the direction of a vector.

Vector subtraction: To get from A to B using vectors, the path is −*a* + *b* or *b* − *a*. Thus the vector $\overrightarrow{AB}$ = *b* − *a*. This general principle should be remembered. It also works with position vectors.

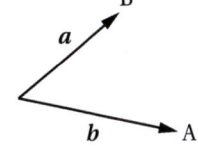

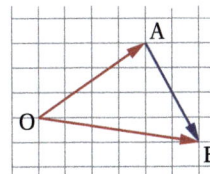

For example, if A is (4, 3) and B is (6, −1) then A and B have position vectors $\binom{4}{3}$ and $\binom{6}{-1}$.

So vector $\overrightarrow{AB}$ = *b* − *a* = $\binom{6}{-1} - \binom{4}{3} = \binom{2}{-4}$.

Alternatively:

$\overrightarrow{OA} = 4i + 3j, \overrightarrow{OB} = 6i - j$

$\overrightarrow{AB} = \overrightarrow{OB} - \overrightarrow{OA}$
$= (6i - j) - (4i + 3j)$
$= (2i - 4j)$

A quadrilateral OABC has points O(0, 0) and A(4, 1).

(a) If $\overrightarrow{AB} = \binom{1}{-2.5}$, find the coordinates of B.

(b) (i) $\overrightarrow{OC} = \binom{-1}{k}$. Find *k* such that $\overrightarrow{CB}$ is parallel to $\overrightarrow{OA}$.

(ii) What is the ratio of the lengths of OA to CB?

(a) $\overrightarrow{OB} = \overrightarrow{OA} + \overrightarrow{AB} = \binom{4}{1} + \binom{1}{-2.5} = \binom{5}{-1.5}$

B = (5, −1.5)

(b) (i) $\overrightarrow{CB} = \overrightarrow{OB} - \overrightarrow{OC} = \binom{5}{-1.5} - \binom{-1}{k}$

$= \binom{6}{-1.5 - k}$

$\binom{6}{-1.5 - k} = m\binom{4}{1}$

So *m* = 1.5 ⇒ −1.5 − *k* = 1.5

k = −3

(ii) OA to CB is 1:1.5

It's possible to become easily confused in questions like this. To avoid confusion, be rigorous with working. In particular, don't mix up the notation for vectors with the notation for coordinates. And always write down "what am I working out, how am I working it out, what is the answer?" In (b)(i), for example:

- I'm working out vector CB
- I need to subtract OC from OB
- ...and I get the answer.

I've introduced letter *m* since parallel vectors are always multiples of each other. Once you know *k* you can work out the coordinates of all the points, and can then answer anything the examiner throws at you!

MATHEMATICS: ANALYSIS AND APPROACHES HL

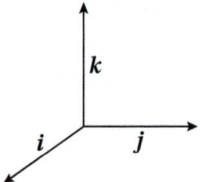

It is always hard to draw accurate diagrams of 3-d vectors. If it *has* to be done, the configuration as shown is often used, but others are possible.

Three-dimensional vectors: Conventional geometry is harder in three dimensions than in two: using vectors, the routines and calculations are pretty much the same in any number of dimensions – even 4! Thus vectors provide us with a powerful framework for solving problems in 3-d geometry.

Magnitude of vectors: The *magnitude* (or length) can be found by using Pythagoras' Theorem in two or three dimensions as appropriate.

The length of $2i + 3j - 5k$ is $\sqrt{2^2 + 3^2 + (-5)^2} = \sqrt{38}$. Use the modulus sign to represent magnitude. Thus, $|p|$ is the magnitude of p.

Collinearity: If three points A, B and C are *collinear* (on the same straight line) then vectors $\overrightarrow{AB}$ and $\overrightarrow{BC}$ must be in the same direction and hence multiples (and $\overrightarrow{AC}$ will be multiples of both of them as well). This provides a simple test for collinearity. Consider the points A(2, 0, 3), B(3, 6, –1) and C(4.5, 15, –7):

$\overrightarrow{AB}$ is $\begin{pmatrix} 1 \\ 6 \\ -4 \end{pmatrix}$ and $\overrightarrow{BC}$ is $\begin{pmatrix} 1.5 \\ 9 \\ -6 \end{pmatrix}$. Thus $\overrightarrow{BC} = 1.5 \times \overrightarrow{AB}$, and therefore A, B and C are collinear.

Unit vectors: Any vector with length 1 is a *unit vector*. The three vectors i, j and k are unit vectors which form the basis of the 3-d coordinate system. As with 2-d vectors, it is easier to use the column format eg: $\begin{pmatrix} 2 \\ 3 \\ 4 \end{pmatrix}$ rather than $2i + 3j + 4k$. To find a unit vector in a particular direction, divide the given vector by its length.

Example: Find the unit vector in the direction $\overrightarrow{AB}$ where A is (4, –1, –2) and B is (6, 0, –4).

Solution: $\overrightarrow{AB} = \begin{pmatrix} 6 \\ 0 \\ -4 \end{pmatrix} - \begin{pmatrix} 4 \\ -1 \\ -2 \end{pmatrix} = \begin{pmatrix} 2 \\ 1 \\ -2 \end{pmatrix}$. Thus $|\overrightarrow{AB}| = \sqrt{2^2 + 1^2 + (-2)^2} = \sqrt{9} = 3$.

The unit vector is therefore $\frac{1}{3}\begin{pmatrix} 2 \\ 1 \\ -2 \end{pmatrix} = \begin{pmatrix} \frac{2}{3} \\ \frac{1}{3} \\ -\frac{2}{3} \end{pmatrix}$.

3.9 Scalar (Dot) Product

Definition: The scalar product is a number which can be calculated from two vectors. On its own it has no real significance, but is used particularly in connection with angles between vectors. The scalar product of two vectors (2-d or 3-d) a and b is defined as: $a \cdot b = |a||b|\cos\theta$ where θ is the angle between the directions of the two vectors. This formula can be read as: "The dot product of vectors a and b = the length of a times the length of b times the cosine of the angle between them." If the vectors are defined in column form, an alternative way of calculating the scalar product is:

$$\begin{pmatrix} a_1 \\ a_2 \\ a_3 \end{pmatrix} \cdot \begin{pmatrix} b_1 \\ b_2 \\ b_3 \end{pmatrix} = a_1 b_1 + a_2 b_2 + a_3 b_3.$$

3. GEOMETRY AND TRIGONOMETRY

Properties of the scalar product: Many of the properties are similar to the algebraic multiplication of numbers $a \times b$.

- $\boldsymbol{a} \cdot \boldsymbol{b} = \boldsymbol{b} \cdot \boldsymbol{a}$
- $\boldsymbol{a} \cdot \boldsymbol{a} = |\boldsymbol{a}^2|$
- $\boldsymbol{a} \cdot (\boldsymbol{b} + \boldsymbol{c}) = \boldsymbol{a} \cdot \boldsymbol{b} + \boldsymbol{a} \cdot \boldsymbol{c}$
- $(m\boldsymbol{a}) \cdot (n\boldsymbol{b}) = mn(\boldsymbol{a} \cdot \boldsymbol{b})$

An important property is that perpendicular vectors have a dot product = 0 (since $\cos 90° = 0$)

If $\boldsymbol{a} \perp \boldsymbol{b}$ then $\boldsymbol{a} \cdot \boldsymbol{b} = 0$ (and vice versa).

Angle between two vectors: The dot product provides a convenient way of calculating the angle between two vectors. This is very common in exams, either as a short question or as part of a longer one. For example, to find the angle between $\boldsymbol{a} = 2\boldsymbol{i} + 3\boldsymbol{j} - \boldsymbol{k}$ and $\boldsymbol{b} = 4\boldsymbol{i} - 2\boldsymbol{j} + 3\boldsymbol{k}$:

$|\boldsymbol{a}|$ is $\sqrt{2^2 + 3^2 + (-1)^2} = \sqrt{14}$; and $|\boldsymbol{b}|$ is $\sqrt{4^2 + (-2)^2 + 3^2} = \sqrt{29}$.

So $\boldsymbol{a} \cdot \boldsymbol{b} = \sqrt{14}\sqrt{29} \cos\theta$. But $\boldsymbol{a} \cdot \boldsymbol{b}$ can also be calculated using column vectors:

$$\boldsymbol{a} \cdot \boldsymbol{b} = \begin{pmatrix} 2 \\ 3 \\ -1 \end{pmatrix} \cdot \begin{pmatrix} 4 \\ -2 \\ 3 \end{pmatrix} = 8 - 6 - 3 = -1$$

So, $\sqrt{14}\sqrt{29} \cos\theta = -1 \Rightarrow \cos\theta = -0.0496$, and finally we get the angle θ to be $\arccos(-0.0496) = 92.8°$. If the question asks for the acute angle between the corresponding *lines*, we must then give the answer 87.2°.

Here's a question using the property that if $\boldsymbol{a} \perp \boldsymbol{b}$ then $\boldsymbol{a} \cdot \boldsymbol{b} = 0$:

Example: If $\boldsymbol{p} = 3\boldsymbol{i} + 2a\boldsymbol{j} - \boldsymbol{k}$ and $\boldsymbol{q} = \boldsymbol{i} + (a - 3)\boldsymbol{j} + (3a - 1)\boldsymbol{k}$, find the possible values of a such that $\boldsymbol{p}$ is perpendicular to $\boldsymbol{q}$.

Solution: $\begin{pmatrix} 3 \\ 2a \\ -1 \end{pmatrix} \cdot \begin{pmatrix} 1 \\ a - 3 \\ 3a - 1 \end{pmatrix} = 0 \Rightarrow 3 + 2a^2 - 6a - 3a + 1 = 0$

$2a^2 - 9a + 4 = 0 \Rightarrow (2a - 1)(a - 4) = 0$

Thus $a = 0.5$ or 4.

MATHEMATICS: ANALYSIS AND APPROACHES HL

> Triangle RST has vertices at R(1, −1, 4), S(2, −1, 0), T(0, 1, 1)
>
> (a) Find angle R.
> (b) Hence find the area of the triangle.

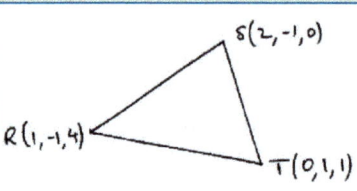

(a) $\vec{RS} = \begin{pmatrix} 1 \\ 0 \\ -4 \end{pmatrix}$; $\vec{RT} = \begin{pmatrix} -1 \\ 2 \\ -3 \end{pmatrix}$;

$\vec{RS} \cdot \vec{RT} = -1 + 0 + 12 = 11$

$\sqrt{17}\sqrt{14} \cos R = 11$

$\cos R = \dfrac{11}{\sqrt{17}\sqrt{14}}$

$R = 44.5°$

(b) Area $= \tfrac{1}{2}$ RS × RT × sin 44.5 = 5.41

In 3-d vector questions, I often find it useful to draw a quick sketch. The sketch won't show points in actual positions, but it helps to envisage what's going on. This becomes an even more vital tool in the lines and planes questions we shall be tackling later in the chapter.

The vector directions are important. Angle R is between vectors $\vec{RS}$ and $\vec{RT}$ so using $\vec{SR}$, for example, will give the wrong answer.

3.10 Vector (Cross) Product

Definition: Unlike the scalar product, the vector product of two vectors is itself a vector. The vector product is written as $a \times b$ and both a and b have to be vectors. If the vectors are written in column form, then the vector product is:

$$a \times b = \begin{pmatrix} a_1 \\ a_2 \\ a_3 \end{pmatrix} \times \begin{pmatrix} b_1 \\ b_2 \\ b_3 \end{pmatrix} = \begin{pmatrix} a_2 b_3 - a_3 b_2 \\ a_3 b_1 - a_1 b_3 \\ a_1 b_2 - a_2 b_1 \end{pmatrix}$$

Look carefully at the patterns to see how to memorise them.

Properties of the vector product:

- $v \times w = -(w \times v)$ because the subtractions will be the other way around.
- $u \times (v + w) = u \times v + u \times w$
- $(kv) \times w = k(v \times w)$
- $v \times v = \mathbf{0}$ where $\mathbf{0}$ represents the zero vector

Applications: The main property of the vector $a \times b$ is that it is perpendicular to both a and b.

> Use the scalar product to test the answer.

For example, if $a \times b = \begin{pmatrix} 2 \\ 1 \\ -3 \end{pmatrix} \times \begin{pmatrix} 0 \\ -2 \\ 4 \end{pmatrix} = \begin{pmatrix} 4 - 6 \\ 0 - 8 \\ (-4) - 0 \end{pmatrix} = \begin{pmatrix} -2 \\ -8 \\ -4 \end{pmatrix}$, then $\begin{pmatrix} -2 \\ -8 \\ -4 \end{pmatrix}$ is perpendicular to a and b. This is particularly useful when finding normal vectors to planes.

Another useful property is that $|a \times b| = |a||b|\sin\theta$ where θ is the angle between the two vectors. This expression is equivalent to the area of the parallelogram which can be formed from the two vectors, and also leads to the formula for the area of a triangle formed

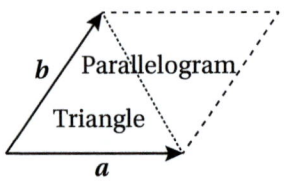

from the two vectors: Area $= \frac{1}{2}|\mathbf{a} \times \mathbf{b}|$. This gives us an alternative method for finding the area of the triangle in the exam-style question in the previous section:

$$\overrightarrow{RS} \times \overrightarrow{RT} = \begin{pmatrix} 1 \\ 0 \\ -4 \end{pmatrix} \times \begin{pmatrix} -1 \\ 2 \\ -3 \end{pmatrix} = \begin{pmatrix} 0-(-8) \\ 4-(-3) \\ 2-0 \end{pmatrix} = \begin{pmatrix} 8 \\ 7 \\ 2 \end{pmatrix}$$

$$|\overrightarrow{RS} \times \overrightarrow{RT}| = \sqrt{8^2 + 7^2 + 2^2} = \sqrt{117}$$

Area of triangle $= \frac{1}{2}\sqrt{117} = 5.41$

3.11 Equations of Lines

Vector equation of a line: Although we can use the Cartesian equation of a line in two dimensions $(y = mx + c)$, in three dimensions it becomes very unwieldy (see below). The *vector equation* of a line is much easier to use, and has the same form in both two and three dimensions.

The line shown in the diagram has a direction given by the vector $\begin{pmatrix} 2 \\ -1 \end{pmatrix}$. We can find the position vector of any point on the line by first going along a vector which **takes** us to the line – say, to the point (2, 4) – and then adding any multiple of the direction vector. This gives us the *vector equation* of the line.

In this case it would be: $\mathbf{r} = \begin{pmatrix} 2 \\ 4 \end{pmatrix} + t\begin{pmatrix} 2 \\ -1 \end{pmatrix}$.

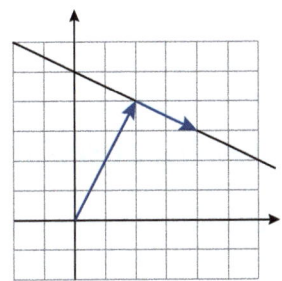

- The $\mathbf{r}$ indicates the position vector of a general point on the line and could also be written as $\begin{pmatrix} x \\ y \end{pmatrix}$.
- The vector $\begin{pmatrix} 2 \\ 4 \end{pmatrix}$ is the position of a point on the line – any other point on the line could have been used.
- t is called the *parameter*. Different values of t give us different points. For example, if $t = 2$, we get the point (6, 2), and every point on the line corresponds to a particular value of t.
- The vector $\begin{pmatrix} 2 \\ -1 \end{pmatrix}$ is the *direction vector* of the line. Other multiples, such as $\begin{pmatrix} 4 \\ -2 \end{pmatrix}$ or $\begin{pmatrix} -2 \\ 1 \end{pmatrix}$ could have been used.

Any letter can be used for the parameter, including Greek letters. λ is often used.

In general, the vector equation of a line is $\mathbf{r} = \mathbf{a} + \lambda \mathbf{b}$ where $\mathbf{a}$ is the position vector of a point on the line and $\mathbf{b}$ is a direction vector. The only difference in 3 dimensions is that $\mathbf{r}$, $\mathbf{a}$ and $\mathbf{b}$ are 3-d vectors.

Example: Find a vector equation of the line passing through (−3, 4, 1) and (2, −2, 2).

Solution: The direction vector of the line is $\begin{pmatrix} 2 \\ -2 \\ 2 \end{pmatrix} - \begin{pmatrix} -3 \\ 4 \\ 1 \end{pmatrix} = \begin{pmatrix} 5 \\ -6 \\ 1 \end{pmatrix}$. We could now use either point as the position vector – I tend to use the one with the simpler numbers if there's a choice.

Thus a vector equation of the line is $\mathbf{r} = \begin{pmatrix} 2 \\ -2 \\ 2 \end{pmatrix} + \lambda \begin{pmatrix} 5 \\ -6 \\ 1 \end{pmatrix}$.

It also follows that a general point on the line could be written as:
$(2 + 5\lambda, -2 - 6\lambda, 2 + \lambda)$.

Cartesian equation of a line: In 3 dimensions, the general form is:

$\frac{x - x_0}{l} = \frac{y - y_0}{m} = \frac{z - z_0}{n}$ where $\begin{pmatrix} l \\ m \\ n \end{pmatrix}$ is the direction vector and (x_0, y_0, z_0) is a point on the line. Since this is the same information as in the vector equation, it is easy to convert from one to the other. The line above therefore has a Cartesian equation $\frac{x + 3}{5} = \frac{y - 4}{-6} = \frac{z - 1}{1}$ (and each of these fractions equals λ).

There are, however, a couple of special cases to be aware of. Consider this Cartesian equation: $\frac{2 - x}{5} = \frac{y}{6} = \frac{z - 1}{1}$. To be able to read off the point and direction vector, it is necessary to rewrite it in the general form. So it becomes $\frac{x - 2}{-5} = \frac{y - 0}{6} = \frac{z - 1}{1}$ and we can see the point is $(2, 0, 1)$. But what if the direction vector has 0 as one of its components? For example, let's try writing down the Cartesian equation of the line through $(1, -1, 3)$ with direction $\begin{pmatrix} 2 \\ 0 \\ 1 \end{pmatrix}$. It should be $\frac{x - 1}{2} = \frac{y + 1}{0} = \frac{z - 3}{1}$ but this would involve division by zero.

Since the *y*-coordinate will **always** be -1, the equation is written as $\frac{x - 1}{2} = \frac{z - 3}{1}, y = -1$.

3.12 Equations of Planes

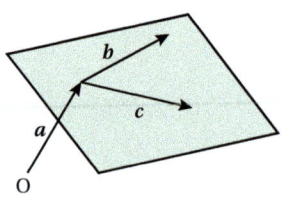

Vector equation of a plane: The vector equation of a plane is similar to that for a line, except that **two** direction vectors are needed to define a plane. The equation will be of the form $\mathbf{r} = \mathbf{a} + \lambda \mathbf{b} + \mu \mathbf{c}$. The problem with this form is that $\mathbf{a}$ can be the position vector of *any* point on the plane, and $\mathbf{b}$ and $\mathbf{c}$ can be any two of the infinite number of directions in the plane. However, it is very easy to find any points on the plane by substituting any values of λ and μ.

> Thus, planes with the same left hand side must have the same normal vector, and are therefore parallel to each other. The larger $|d|$ is, the further away the plane is from the origin.

Cartesian equation of a plane: Unlike the line, the Cartesian equation of the plane is much easier to use. It is always of the form $ax + by + cz = d$. Although a plane contains an infinite number of direction vectors, only **one** direction can be perpendicular to a plane. A vector in this direction is called a *normal vector*, and it turns out that the plane $ax + by + cz = d$ has a normal vector $\begin{pmatrix} a \\ b \\ c \end{pmatrix}$. The vector product gives us a useful way to find the Cartesian equation of a plane given three points in the plane.

1. Use the three points to find two direction vectors
2. Use the cross product to find a vector perpendicular to these – this will be the normal vector for the plane.
3. Write down the LHS of the equation.
4. Substitute any of the three points to work out the RHS.

Let's follow this method to find the equation of the plane containing the points A(2, 4, 0), B(−1, 0, 5) and C(4, 2, −2).

1. Trying to end up with as many positive values as possible, I choose to find directions BA and CA.

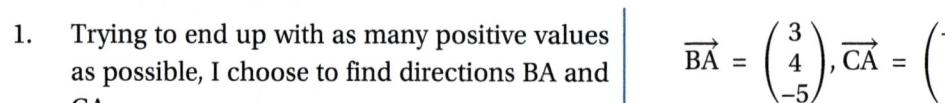

3. GEOMETRY AND TRIGONOMETRY

2. Now find the cross product. Since this gives a direction vector it can be simplified if possible.

$$\overrightarrow{BA} \times \overrightarrow{CA} = \begin{pmatrix} 18 \\ 4 \\ 14 \end{pmatrix} \text{ or } \begin{pmatrix} 9 \\ 2 \\ 7 \end{pmatrix}$$

3. This gives the LHS of the equation of the plane.

$$9x + 2y + 7z = d$$

4. Substitute any of the points (B looks the simplest) to complete the equation.

$$9x + 2y + 7z = 26$$

And it's then a good idea to check the equation by substituting either of the other two points. For example, substituting A gives $18 + 8 + 0 = 26$.

Alternative form for the Cartesian equation: The general form of the equation $ax + by + cz = d$ can be written using the dot product as $\begin{pmatrix} x \\ y \\ z \end{pmatrix} \cdot \begin{pmatrix} a \\ b \\ c \end{pmatrix} = d$ or $\boldsymbol{r \cdot n} = d$, where $\boldsymbol{n}$ is the normal vector. It can also be shown that $d = \boldsymbol{a \cdot n}$ where $\boldsymbol{a}$ is a fixed point on the plane. Thus we get the alternative form $\boldsymbol{r \cdot n} = \boldsymbol{a \cdot n}$ which **looks** like a vector equation but is in fact an alternative expression for the Cartesian equation.

Show that the line $\boldsymbol{r} = \begin{pmatrix} 2 \\ 1 \\ 3 \end{pmatrix} + \lambda \begin{pmatrix} 4 \\ 3 \\ 1 \end{pmatrix}$ is parallel to the plane $x - 2y + 2z = 5$.

Direction vector of line is $\begin{pmatrix} 4 \\ 3 \\ 1 \end{pmatrix}$ Normal vector to plane is $\begin{pmatrix} 1 \\ -2 \\ 2 \end{pmatrix}$ $\begin{pmatrix} 4 \\ 3 \\ 1 \end{pmatrix} \cdot \begin{pmatrix} 1 \\ -2 \\ 2 \end{pmatrix} = 4 - 6 + 2 = 0$ Line and normal are perpendicular, so the line must be parallel to the plane.	It very often helps to draw a quick sketch when dealing with vector questions – the sketch helps you decide which techniques to use. 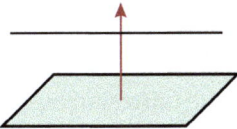 So you can see that for a line to be parallel to a plane, it must also be perpendicular to the normal vector of the plane. This implies we put the dot product to work.

The following example uses the same principle as the question above but is worded very differently.

Example: The plane $4x + y - 3z = 6$ contains the line $x - 1 = \dfrac{y-5}{2} = \dfrac{z-1}{k}$.

Find the value of k.

Solution: The direction of the line is $\begin{pmatrix} 1 \\ 2 \\ k \end{pmatrix}$ so $\begin{pmatrix} 1 \\ 2 \\ k \end{pmatrix} \cdot \begin{pmatrix} 4 \\ 1 \\ -3 \end{pmatrix} = 0$.

$4 + 2 - 3k = 0$ and this gives $k = 2$.

MATHEMATICS: ANALYSIS AND APPROACHES HL

Equations of Lines and Planes – Summary

	Line	Plane
Cartesian Equation	$\dfrac{x-a}{l} = \dfrac{y-b}{m} = \dfrac{z-c}{n}$	$ax + by + cz = d$
Notes	Not as easy to use as the vector equation of a 3d line; convert if necessary.	$\begin{pmatrix} a \\ b \\ c \end{pmatrix}$ is the normal vector.
Vector Equation	$r = \begin{pmatrix} a \\ b \\ c \end{pmatrix} + \lambda \begin{pmatrix} l \\ m \\ n \end{pmatrix}$	$r = \begin{pmatrix} f \\ g \\ h \end{pmatrix} + \lambda \begin{pmatrix} l \\ m \\ n \end{pmatrix} + \mu \begin{pmatrix} p \\ q \\ r \end{pmatrix}$
Notes	The a, b, c, l, m, n are the same as in the Cartesian equation.	Two direction vectors are required to define the plane. Less useful than the Cartesian equation.

Answers

1. $r = i + 3j + 2k + \lambda(i - 3j - 4k)$
2. $52.7°$
3. 11.8
4. $5x + 3y - z = 12$
5. $r = 10i + 11j + k + \lambda(5i + 3j - k)$
6. $r = a + \lambda b + \mu c$ where a is the position vector of any of the points B, C, D, and b and c are any two of $2i - 2j - 4k$, $8i + 11j + 3k$, $6i + 13j - k$
7. $\dfrac{x-1}{3} = \dfrac{y-3}{-5}, z = 2$

Equations of Lines and Planes: Practice Exercise

This exercise contains a variety of questions involving the methods and techniques covered so far in this chapter. They all use the four points A(1, 3, 2), B(2, 0, −2), C(4, −2, 2) and D(10, 11, 1).

1. Write down the vector equation of line (AB).
2. Find the acute angle between lines (AB) and (AC).
3. Find the area of triangle ABC.
4. Work out the Cartesian equation of plane ABC.
5. Find the equation of the line which is normal to plane ABC and which goes through the point D.
6. Find a vector equation for plane BCD.
7. Find a Cartesian equation for line (AC).

3.13 Intersections

You must learn a "library" of techniques which will help you to answer the majority of vector geometry questions. You have already seen some of them in the preceding sections; the next group involve intersections between lines and lines, lines and planes, and planes and planes. Throughout, we are working in 3 dimensions.

Intersection of two lines: There are three possible orientations of two lines:

- The two lines intersect
- The two lines are parallel
- The two lines are not parallel and do not intersect (*skew* lines).

We can easily recognise two parallel lines – they have the same direction vector (or they are multiples of each other). In all other cases, we begin by putting the vector equations equal to each other (if a line has a Cartesian equation, convert it).

Example: Find if lines $r = -i + 4j + \lambda(3i - 2j + k)$ and $\dfrac{x-4}{2} = \dfrac{y-4}{-3} = \dfrac{z+1}{2}$ meet. If they do, find the point of intersection.

3. GEOMETRY AND TRIGONOMETRY

Stage 1: Write the two lines in "condensed vector" form *(my term)*.

$$\text{Line 1 is } \begin{pmatrix} -1 + 3\lambda \\ 4 - 2\lambda \\ \lambda \end{pmatrix} \quad \text{Line 2 is } \begin{pmatrix} 4 + 2\mu \\ 4 - 3\mu \\ -1 + 2\mu \end{pmatrix}$$

Make sure you understand how both lines have been converted to this form. Also note that vector form can be called "parametric form."

Stage 2: Equate the *x*, *y* and *z* components to find where they have the same *x*, *y* and *z* coordinates.

$$-1 + 3\lambda = 4 + 2\mu$$
$$4 - 2\lambda = 4 - 3\mu$$
$$\lambda = -1 + 2\mu$$

Three equations in two unknowns? Two of the equations will be enough to find λ and μ; then, if those values do not satisfy the third equation, the lines do not intersect. Let's try it...

Stage 3: Solve two of the equations (whichever look easiest). Solving the second and third simultaneously gives $\mu = 2$, $\lambda = 3$. Now try these in the first equation: LHS = $-1 + 3 \times 3 = 8$. RHS = $4 + 2 \times 2 = 8$. So the lines *do* intersect.

Stage 4: To find the point of intersection, simply substitute λ or μ into the equations of the lines. We get (8, -2, 3) as the point of intersection.

It is only necessary to substitute one of the values, but do it with both and you have a good check for your answer.

For a bit of practice, work out if the following pairs of lines intersect; if not, say whether they are skew or parallel:

(a) $\frac{x-1}{2} = y + 2 = \frac{z-3}{-1}, \mathbf{r} = \begin{pmatrix} -1 \\ 3 \\ 7 \end{pmatrix} + \lambda \begin{pmatrix} -2 \\ 1 \\ 2 \end{pmatrix}$.

(b) $\mathbf{r} = 3\mathbf{i} + 2\mathbf{j} + 4\mathbf{k} + \lambda(\mathbf{i} + 2\mathbf{j} - \mathbf{k}), \frac{x-2}{3} = \frac{y-4}{6} = \frac{z+1}{-3}$.

(c) $\frac{y-4}{2} = \frac{z}{-1}, x = 1, \mathbf{r} = \begin{pmatrix} 1 \\ 2 \\ 1 \end{pmatrix} + \lambda \begin{pmatrix} 4 \\ 2 \\ -3 \end{pmatrix}$.

(d) $\mathbf{r} = \begin{pmatrix} 2 \\ -3 \\ -1 \end{pmatrix} + \lambda \begin{pmatrix} -1 \\ 3 \\ 2 \end{pmatrix}, \mathbf{r} = \begin{pmatrix} 3 \\ 7 \\ 6 \end{pmatrix} + \mu \begin{pmatrix} 3 \\ 4 \\ 2 \end{pmatrix}$.

Answers
(a) Meet at (5, 0, 1)
(b) Parallel
(c) Meet at (1, 2, 1)
(d) Skew

Intersection of a line and a plane: Again, there are three possibilities:

- the line intersects the plane at a single point;
- the line is parallel to the plane;
- the line lies in the plane.

We're going to form an equation in the parameter, λ. If there is just one (*unique*) solution, we've got the point of intersection. If the solution "goes wrong", there will either be no solutions (line parallel to the plane) or an infinite number of solutions (line lies in the plane). Let's see an example of each: the "setup" is the same in each case, and consists of writing the equation of the line in "condensed vector" form then substituting each of the three components into the (Cartesian) equation of the plane.

Example: Find where the plane $2x + 3y + z = 7$ and line $\mathbf{r} = \begin{pmatrix} 2 \\ -1 \\ -2 \end{pmatrix} + \lambda \begin{pmatrix} 1 \\ 0 \\ 2 \end{pmatrix}$ meet.

Solution: Substituting $x = 2 + \lambda$, $y = -1$ and $z = -2 + 2\lambda$ into the equation of the plane gives: $2(2 + \lambda) + 3(-1) + (-2 + 2\lambda) = 7$, which solves to give $\lambda = 2$. Put this back into the equation of the line to get the point of intersection as (4, -1, 2).

MATHEMATICS: ANALYSIS AND APPROACHES HL

Example: Show that the line $r = \begin{pmatrix} 0 \\ 4 \\ 2 \end{pmatrix} + \lambda \begin{pmatrix} 3 \\ -2 \\ -1 \end{pmatrix}$ lies in the plane $x + 2y - z = 6$.

Solution: This time substitution gives us $(3\lambda) + 2(4 - 2\lambda) - (2 - \lambda) = 6$ which simplifies to $0\lambda + 6 = 6$, which is clearly true for all values of λ. Thus all points on the line intersect with the plane, and hence it must lie in the plane.

Example: Show that line $r = \begin{pmatrix} 2 \\ 3 \\ -1 \end{pmatrix} + \lambda \begin{pmatrix} 6 \\ 3 \\ 0 \end{pmatrix}$ is parallel to the plane $r \cdot \begin{pmatrix} 1 \\ -2 \\ 3 \end{pmatrix} = 5$.

Solution: The plane has been written in the $r \cdot n = d$ format but is exactly the same as $x - 2y + 3z = 5$. Substituting, we get $(2 + 6\lambda) - 2(3 + 3\lambda) - 3 = 5$ which simplifies to $-7 = 5$. Clearly this is never true; there are no intersections, so the line must be parallel to the plane.

Intersection of two planes: Two planes are either parallel (in which case they have the same or parallel normal vectors) or they intersect along a line. How do we find the equation of the line of intersection? The steps are:

- Use both equations to eliminate one letter, say z.
- Do the same again to eliminate another letter, say y.
- Make the third letter (x) the subject of both equations.
- The Cartesian equation of the line can now be written down.

Example: Find the equation of the line of intersection of the planes:

$$x - y + 2z = 3 \text{ and } 2x + y + 3z = 1.$$

Solution: Eliminating y from the equations gives $3x + 5z = 4$.

Eliminating x from the equations gives $3y - z = -5$.

Make z the subject of both to get $z = \dfrac{-3x + 4}{5}$ and $z = 3y + 5$.

So the equation of the line of intersection is

$$\dfrac{-3x + 4}{5} = 3y + 5 = z \text{ or } \dfrac{x - \frac{4}{3}}{-\frac{5}{3}} = \dfrac{y + \frac{5}{3}}{\frac{1}{3}} = \dfrac{z - 0}{1}$$

One way to check this is to see if the point on the line ie $\left(\frac{4}{3}, -\frac{5}{3}, 0\right)$ lies on both planes.

> Although the vector equation of a line is easier to use, this method leads straight to the Cartesian equation. No particular form was specified in the question.

Intersection of three planes: Use the row reduction techniques discussed on page 31. If you end up with three zeroes on the bottom line then there will be either 0 solutions (the planes form a prism with a common axis), or ∞ solutions (a *sheaf* of planes meeting along a common line). You can tell which is which by looking along the bottom row to the number on the right hand side - this will be 0 for a sheaf, any other number for a prism.

Consider the intersections of the following planes: $\begin{cases} x + y + z = 5 \\ 3x - y + 2z = 11 \\ 5x + ay + bz = q \end{cases}$

Show that when:

- $a = -3, b = 3, q = 17$, the planes meet in a line (a sheaf);
- $a = -3, b = 3, q = 16$, the planes make a prism;
- $a = 5, b = 5, q = 25$, two planes are parallel;
- $a = -3, b = 2, q = 17$, the planes meet at the point (4, 1, 0).

3.14 Angles in Three Dimensions

On page 85 you were shown how to find the angle between two vectors using the scalar product. When dealing with vectors, you will use this technique whenever you need to find an angle. It is then only necessary to decide which two vectors to use.

Angle between two lines: The angle between two lines is the same as the angle between their direction vectors. For example:

Lines $r = \begin{pmatrix} 1 \\ -1 \\ 3 \end{pmatrix} + \lambda \begin{pmatrix} 2 \\ 1 \\ 4 \end{pmatrix}$ and $r = \begin{pmatrix} 0 \\ 4 \\ 1 \end{pmatrix} + \mu \begin{pmatrix} -1 \\ 2 \\ 1 \end{pmatrix}$ have direction vectors $\begin{pmatrix} 2 \\ 1 \\ 4 \end{pmatrix}$ and $\begin{pmatrix} -1 \\ 2 \\ 1 \end{pmatrix}$.

The dot product method gives the angle between these directions, and hence between the lines, as 69.1°.

Angle between a line and a plane: A plane has no single direction, but we can find the angle between the *normal* vector (which has a unique direction) and the direction of the line. Subtracting this from 90° gives the angle between the line and the plane.

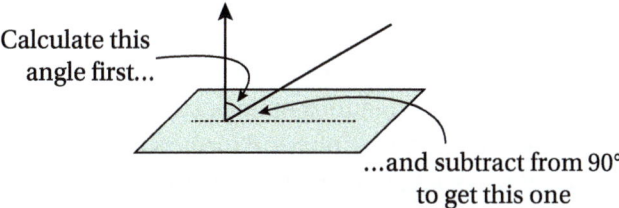

Calculate this angle first...

...and subtract from 90° to get this one

Π_1 is the plane $2x - y + 3z = 8$; Π_2 is the plane $4x - 2y = 10$. Points R and S have coordinates (4, −1, 4) and (3, 2, 0) respectively.

  (i) Find the angle between Π_1 and Π_2.

  (ii) Find the equation of plane Π_3 which contains point R and is perpendicular to both planes Π_1 and Π_2.

  (iii) Find the angle between line RS and plane Π_3.

(i) The normal vectors are $\begin{pmatrix} 2 \\ -1 \\ 3 \end{pmatrix}$ and $\begin{pmatrix} 4 \\ -2 \\ 0 \end{pmatrix}$.

$\begin{pmatrix} 2 \\ -1 \\ 3 \end{pmatrix} \cdot \begin{pmatrix} 4 \\ -2 \\ 0 \end{pmatrix} = 8 + 2 + 0 = 10$

$10 = \sqrt{14}\sqrt{20}\cos\theta$

$\theta = \arccos\left(\dfrac{10}{\sqrt{14}\sqrt{20}}\right) = 53.3°$

(i) Angle between two planes is the angle between their normal vectors, which can be read off from their equations.

(ii) $\begin{pmatrix} 2 \\ -1 \\ 3 \end{pmatrix} \times \begin{pmatrix} 4 \\ -2 \\ 0 \end{pmatrix} = \begin{pmatrix} 6 \\ 12 \\ 0 \end{pmatrix}$ so a normal vector to Π_3 is $\begin{pmatrix} 1 \\ 2 \\ 0 \end{pmatrix}$.

Π_3 is $x + 2y = d$.

Substitute (4, −1, 4) to get $x + 2y = 2$

(ii) If a plane is perpendicular to two others, then its normal vector is perpendicular to both of the other normals – this is a job for the vector product.

MATHEMATICS: ANALYSIS AND APPROACHES HL

> (iii) RS has direction $\begin{pmatrix}3\\2\\0\end{pmatrix} - \begin{pmatrix}4\\-1\\4\end{pmatrix} = \begin{pmatrix}-1\\3\\-4\end{pmatrix}$
>
> $\begin{pmatrix}-1\\3\\-4\end{pmatrix} \cdot \begin{pmatrix}1\\2\\0\end{pmatrix} = -1 + 6 = 5$
>
> $5 = \sqrt{26}\sqrt{5}\cos\theta$ where θ is angle between RS and normal
>
> $\theta = \arccos\left(\dfrac{5}{\sqrt{26}\sqrt{5}}\right) = 64.0°$
>
> Angle between RS and Π_3 = 90 − 64.0 = 26.0°

(iii) The important point here is the necessity of showing clear working. If I had just used angle θ without saying what it represented, I would probably confuse myself as well as the examiner! A few words here and there to explain what you are doing are like gold dust to an examiner.

Angle between two planes: The angle between two planes is the same as the angle between their normal vectors.

Miscellaneous Exam-Style Questions

I've given short answers here, but you'll find fully worked answers on the website.

To answer these questions, you need (a) lots of practice with the whole range of vector methods – basic vector operations, equations of lines and planes, angles, intersections – and (b) a good mental image, often taking physical form as a sketch, of what the question is about. If there are several parts, add information to your sketch as you work through the question.

1. Find the point A on the line $r = 3i + j + \lambda(2i + j + 2k)$, $\lambda \in \mathbb{Z}$, which is 5 units away from point B(1, 2, −1).

2. Lines L and M have equations $\dfrac{x-1}{2} = \dfrac{y}{2}, z = 1$ and $\dfrac{x+1}{3} = \dfrac{y-2}{-1} = z$ respectively. Find:

    (i) the point of intersection of L and M.

    (ii) the Cartesian equation of the plane containing both lines.

3. (a) Show that the points O(0, 0, 0), A(6, 0, 0), B(6, −√24, √12), C(0, −√24, √12) form a square.

    (b) Find a Cartesian equation of the plane Π containing OABC

    (c) Find a vector equation for the line which is perpendicular to Π and which passes through the midpoint, M, of OB.

4. A = (4, 3, 3) and B = (5, 0, −5). Π is the plane $x − y + 2z = −5$.

    (a) Show that B lies on Π.

    (b) Find a vector equation for line AB.

    (c) Write down a vector equation for the line L through A which is perpendicular to Π.

    (d) Find C, the point of intersection of L and Π.

    (e) Hence show that the area of triangle ABC is $5\sqrt{12}$.

    (f) Find the point A′ the reflection of A in Π.

5. Find the acute angle between the planes $x − z = 2$ and $2x + 3y + z = 4$.

Answers

1. (5, 2, 2)
2. (i) (2, 1, 1)
   (ii) $x − y − 4z = −3$
3. (b) $y + \sqrt{2}\,z = 0$
   (c) $r = \begin{pmatrix}3\\-\sqrt{6}\\\sqrt{3}\end{pmatrix} + \lambda\begin{pmatrix}0\\1\\\sqrt{2}\end{pmatrix}$
4. (b) $r = \begin{pmatrix}4\\3\\3\end{pmatrix} + \lambda\begin{pmatrix}1\\-3\\-8\end{pmatrix}$
   (c) $r = \begin{pmatrix}4\\3\\3\end{pmatrix} + \lambda\begin{pmatrix}1\\-1\\2\end{pmatrix}$
   (d) (2, 5, −1)
   (e) $\tfrac{1}{2}\sqrt{50}\sqrt{24} = 5\sqrt{12}$
   (f) (0, 7, −5)
5. 79.1°
6. (a) $a = −1, b = 2, c = 4$
   (b) 3

6. (a) Given that $u = i + 2j - 2k$, $v = \begin{pmatrix} a \\ b \\ 0 \end{pmatrix}$ and $u \times v = \begin{pmatrix} 4 \\ b \\ c \end{pmatrix}$, find the values of a, b and c.

   (b) Given that u and v are the position vectors of points U and V, find the area of triangle OUV, where O is the origin.

Kinematics: Vectors can be used to solve simple kinematics questions where the object is moving in a straight line at constant speed. Suppose, for example, the position of a model plane is given by $r = \begin{pmatrix} 1 \\ 4 \\ 0 \end{pmatrix} + t \begin{pmatrix} 3 \\ -1 \\ 2 \end{pmatrix}$ where t is in seconds and distances in m. The initial position is $(1, 4, 0)$, and the height at time t is $2t$. The change in position each second, ie the velocity, is the direction vector, and the speed is the magnitude of this vector which is $\sqrt{14}$ ms^{-1}. A video on the website shows an example of a kinematics exam question.

Geometry and Trigonometry: Long Answer Questions

Starting on the next page is a selection of Paper 2 style exam questions. The answers are given here, but full working may be found on the website (www.peakib.com).

MATHEMATICS: ANALYSIS AND APPROACHES HL

1. (a) (i) Sketch the graph of $y = \sin 2x$, $0 \leq x \leq \pi$.

   (ii) Hence sketch the curve of $y = \text{cosec } 2x$, $0 \leq x \leq \pi$. Show the coordinates of any local maximum and minimum points, and the equations of any asymptotes.

   (b) Show that $\tan x + \cot x = 2\text{cosec } 2x$.

   (c) Find the solution of $2\text{cosec } 2x + 1 = 3\tan x$, $0 \leq x \leq \frac{\pi}{2}$.

Answers:

(a) (i)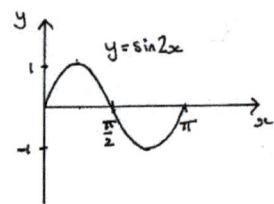

(ii)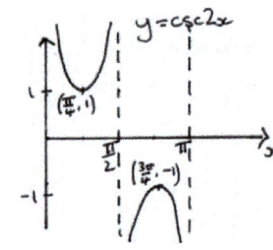

Asymptotes at $x = 0$, $x = \frac{\pi}{2}$, $x = \frac{\pi}{4}$

(c) $x = \frac{\pi}{4}$

---

2. The diagram shows a quadrilateral ABCD. Angles A and C are obtuse.

   (a) Use the cosine rule to show that $BD = \sqrt{52 - 48\cos x}$

   (b) Use the sine rule to find another expression for BD in terms of x.

   (c) By equating the expressions for BD, find x.

   (d) Hence find BD.

   (e) Find y

   (f) Hence find the area of quadrilateral ABCD.

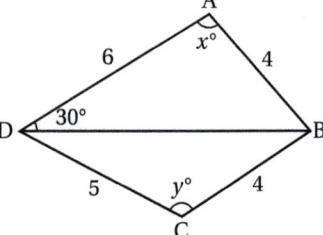

Answers:

(b) $BD = 8\sin x$

(c) $x = 101.4°$

(d) $BD = 7.84$

(e) $y = 120.8°$

(f) 20.4

3. GEOMETRY AND TRIGONOMETRY

3. Let A = (2, -1, 0), B = (3, 0, 1) and C = (1, -1.5, 0.5)

   (a) Determine the Cartesian equation of the plane Π containing ABC.

   (b) Find the area of triangle ABC.

   (c) (i) Show that vector OC is perpendicular to Π, where O is the origin.

   (ii) Hence calculate the distance of Π from the origin.

   (d) Calculate the volume of pyramid OABC, giving your answer as a simplified fraction.

   (e) Calculate the angle OB makes with Π.

Answers:

(a) $2x - 3y + z = 7$

(b) $\frac{1}{2}\sqrt{3.5}$

(c) (ii) $\sqrt{3.5}$

(d) $\frac{7}{12}$

(e) $36.3°$

4. (a) (i) Prove that $\frac{\cos x}{1 + \sin x} + \frac{1 + \sin x}{\cos x} \equiv 2\sec x$, $x \neq (2n+1)\frac{\pi}{2}$, $n \in \mathbb{N}$

   (ii) Explain the restriction on the values of x

   (iii) Solve $\frac{\cos x}{1 + \sin x} + \frac{1 + \sin x}{\cos x} = \frac{4}{\sqrt{3}}$ for $0 \leq x \leq 2\pi$

   (b) (i) Sketch the graph of $y = \sin(2|x| + \pi)$ for $-\pi \leq x \leq \pi$

   (ii) Hence state the values of k such that $\sin(2|x| + \pi) = k$, $-\pi \leq x \leq \pi$ has exactly two solutions.

Answers:

(a) (ii) Those values of x result in division by 0, hence they are vertical asymptotes.

(iii) $x = \frac{\pi}{6}, \frac{11\pi}{6}$

(b) (i)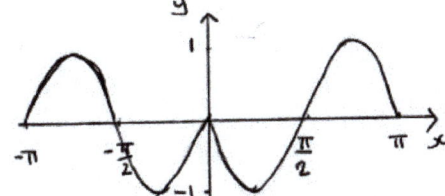

(ii) $k = -1$ or 1

5. Two perpendicular lines l_1 and l_2 have equations $r = \begin{pmatrix} 11 \\ 2 \\ 17 \end{pmatrix} + \lambda \begin{pmatrix} -2 \\ 1 \\ -4 \end{pmatrix}$ and $r = \begin{pmatrix} -5 \\ 11 \\ a \end{pmatrix} + \mu \begin{pmatrix} b \\ 2 \\ 2 \end{pmatrix}$ respectively.

(a) Show that $b = -3$.

Given that l_1 and l_2 intersect, find

(b) (i) The value of a

   (ii) The point of intersection.

(c) Show that point P(9, 3, 13) lies on l_1

A circle with centre C on line l_2 cuts l_1 at points P and Q.

(d) Find the position vector of Q.

Given C is the point $(m, 9, -1)$

(e) (i) Find the value of m.

   (ii) Find the area of triangle CPQ.

___

Answers:

(b) (i) $a = 1$

   (ii) $(1, 7, -3)$

(d) $\begin{pmatrix} -7 \\ 11 \\ -19 \end{pmatrix}$

(e) (i) $m = -2$

   (ii) 75.6

4. STATISTICS AND PROBABILITY

Chapter 4: STATISTICS AND PROBABILITY

4.1 Definitions

A *population* is a set from which statistics are drawn. A *sample* is a subset drawn from the population. In a random sample, every member of the population is equally likely to be chosen. There are several different *sampling techniques*, each with advantages and disadvantages. A sampling technique may introduce *bias* – for example, selecting a sample of people in a shopping centre at 11am will not include those who are at work.

Sample statistics (such as the mean) can be used to estimate population statistics. *Discrete* data are restricted to certain values only (often integers) whereas *continuous* data can take any values. The *frequency* is the number of times a particular value occurs. When collecting data, some items may appear to be very extreme when compared to the rest of the data. Such items are called *outliers* and consideration must be given to whether they *could* be valid, or whether they are incorrect; and also how to deal with them.

Numerical data is usually collected into a *frequency table* and can then be split into *groups* or *classes*. The *boundaries* of the classes must be dealt with carefully, especially for continuous data. Consider a table of weights which begins like this:

Examples of populations:
People who live in Europe
People who drive
Apples grown in France
Cars made in 2002

Examples of discrete data:
Shoe sizes
Goals scored by a team
Number of chocolates in a box

Examples of continuous data
Weights of people
Athletes' times to run 100 m
Heights of mountains

Weight, w (kg)	Frequency
0–10	4
10–20	12
20–30	18

Into which class would a weight of 10kg be put? It would be better if the first group were labelled $0 \leq w < 10$ and the second $10 \leq w < 20$, then 10 would fall into the second group. The *interval width* in this case is 10, and the *mid-interval value* of the first group is 5 and so on. Data can be appreciated more when displayed in a diagram and the *frequency histogram* is the simplest way to display grouped data. A frequency histogram (often called a *bar chart*) uses equal class intervals.

4.2 Averages

Properly called "measures of central tendency", there are three types of average you need to know: mean, median and mode. An average is a single statistic which can be used to represent a whole group, although this isn't true of the mode which merely tells us the "most popular" value.

MATHEMATICS: ANALYSIS AND APPROACHES HL

The mean: To calculate the mean, add all the numbers together and divide by the number of values, n. So mean $= \frac{\sum x_i}{n}$, where the separate values are x_1, x_2, x_3 and so on. The symbol for sample mean is $\bar{x}$. Note that $n\bar{x} = \sum x_i$.

Example: In 9 games I have scored a mean of 12.8 points. In the 10th game I score 16 points – what is my new mean?

Solution: The total score in the first 9 games is $9 \times 12.8 = 115.2$. My new total in 10 games is $115.2 + 16 = 131.2$, so my new mean is $\frac{131.2}{10} = 13.12$.

Here's a similar question:

> 100 people are staying at a hotel: 68 are men and 32 women. The men have a mean height of 1.75 m and the women have a mean height of 1.64 m. Find the mean height of the 100 people.

If you can't get the answer, you'll find full working on the website.

The answer is 1.71 m.

If the data is in a frequency table – such as the one below showing how many pupils were absent during a month – then the total value is calculated by multiplying each value by its frequency and summing the results.

Pupils absent (x)	No of days (f)	fx
0	20	0
1	4	4
2	3	6
3	3	9
TOTAL	30	19

There were a total of 19 days absence over a period of 30 days. So the mean number of absences per day was $\frac{19}{30} = 0.63$. (It is a common mistake to divide 19 by 4, the number of classes).

If the data is presented in a *grouped frequency table*, the same procedure is followed except that the mid-interval value of each group is used to represent the x value for each group. This means that the *actual* data values are unknown and in this case the mean is only an estimate.

Weight of apples (w)	No of apples (f)	Mid interval	fx
$20 \leq w < 25$	12	22.5	270
$25 \leq w < 30$	20	27.5	550
$30 \leq w < 35$	25	32.5	812.5
$35 \leq w < 40$	17	37.5	637.5
TOTAL	74		2270

Always check if the answer is "reasonable." Look at the distribution of weights – does 30.7 look like the mean?

The estimated mean weight of an apple is $\frac{2270}{74} = 30.7$ g.

4. STATISTICS AND PROBABILITY

> The table shows the scores of competitors in a competition.
>
> | Score | 10 | 20 | 30 | 40 | 50 |
> |---|---|---|---|---|---|
> | Number of competitors with this score | 1 | 2 | 5 | k | 3 |
>
> The mean score is 34. Find the value of k.
>
> | | |
> |---|---|
> | Total score = $10 + 40 + 150 + 40k + 150 = 350 + 40k$
 Number of competitors = $1 + 2 + 5 + k + 3 = 11 + k$
 Mean is $\dfrac{350 + 40k}{11 + k} = 34$
 $350 + 40k = 34(11 + k)$
 $350 + 40k = 374 + 34k$
 $6k = 24$
 $k = 4$ | *At first sight you may be puzzled as to how to tackle a question like this. But, knowing the formula for the mean, just carry on and see what happens!* |

Your GDC will calculate the mean of a set of numbers within its statistical functions. But be careful: if you want the mean of a frequency table you will need two lists (the values and the frequencies), and the GDC will need to know that the second list contains the frequencies. If you're uncertain about this, try it with the table above.

Median: If a set of values is listed in order, the middle value is the *median*. It is another type of average: there are as many values above the median as below it. Unlike the mean, it is unaffected by particularly large or small values. In the following list there are 15 values so the 8th is the middle one (7 below it, 7 above it).

In general, if there are n values, the median is in the $\dfrac{n+1}{2}$th position.

$$1 \ 1 \ 3 \ 5 \ 6 \ 6 \ 6 \ \underline{7} \ 7 \ 9 \ 10 \ 10 \ 12 \ 15 \ 18 \rightarrow \text{median} = 7$$

If there is an even number of values, find the mean of the middle two to calculate the median.

$$24 \ 26 \ 27 \ \underline{27 \ 29} \ 30 \ 30 \ 33 \rightarrow \text{median} = 28$$

If the data is in the form of a frequency table, then the calculation depends on whether it is discrete or continuous.

Discrete distribution

x	1	2	3	4	5	6
f	4	11	17	25	14	4

 Beware! If you enter a grouped frequency table, you will **not** get correct values for the median and the quartiles.

There are 75 values, so the median will be the 38th. The first 4 values are 1s, the next 11 are 2s, making 15 values so far. Another 17 are 3s making 32 values. So the 38th value must be in the next box, and thus the median is 4.

Continuous distribution

x	0 –	5 –	10 –	15 –	20 –	25 – 30
f	4	11	17	25	14	4

This time, the values are spread throughout each class, so the 38th value will be the 6th in the class 15–20.

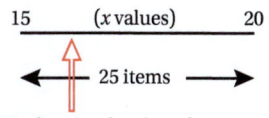

Median is 6th value of 25 items

101

Interpolating, median = $15 + \frac{6}{25} \times 5 = 16.2$

Cumulative frequency tables: It is slightly easier to estimate the median from a frequency table if it is first converted into a *cumulative frequency table*. Whether the data is discrete or continuous, the method is the same. Each value of cumulative frequency measures how many x values there are in total up to that point. The two tables above convert into the following:

x	≤ 1	≤ 2	≤ 3	≤ 4	≤ 5	≤ 6
Cumul. f	4	15	32	57	71	75

In the first table we can see that there are 32 values up to 3, so the 38th value must be contained in the next group and is 4.

x	<5	<10	<15	<20	<25	<30
Cumul. f	4	15	32	57	71	75

> Note that in the conversion of the grouped frequency table, the "up to" points are the *top* of each group.

In the second table we have to recalculate the fact that there are 25 values in the group 15–20, and then go on to the calculation shown above. The advantage here is not so great, but we can go one stage further and draw a cumulative frequency graph to help us.

The points in the table are plotted and are joined either by straight lines or a smooth curve. To find the median, a line is drawn to the right from 38 (the middle value of the distribution) and down to the x axis.

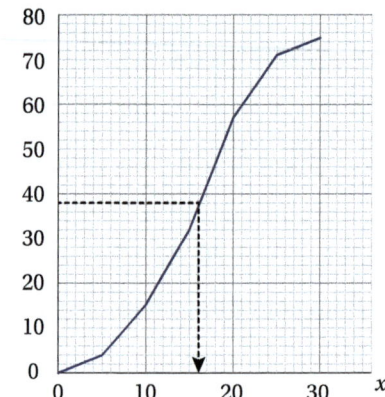

Cumulative frequency

The median can be seen to be about 16.

Quartiles: 50% of the population lie above the median, 50% below. We can also divide the population into *quartiles*: 25% lie below the first quartile, 50% below the second (which is also the median) 75% below the third quartile. There are 75 results in the previous table, so the first quartile will be the 19th result. Looking at the graph, this gives the first quartile as 11 and the third quartile (the 57th result) as 20. Similarly, the distribution can be divided into 100 parts knows as *percentiles*. "Your test result is in the top 5 percentiles of the population" means that at least 95% of people scored worse than you did.

> There are various methods for calculating quartiles of a discrete distribution. For the exam, you will be expected either to use a graph or your GDC

Mode: The mode, or modal value, is simply the value which occurs the most often in a frequency distribution. In other words, the value with the greatest frequency. In a grouped frequency distribution, the group with the greatest frequency is called the *modal class*. Note that there can be more than one modal value or class.

Puzzle: Can you find five numbers such that mode < median < mean? And can you find five numbers such that mode < mean < median?

Solution: There are lots of possibilities, such as 2, 2, 5, 10, 12 and 2, 2, 6, 7, 8.

4.3 Measures of spread

The mean and the median give an indication of the "centre" of the distribution. The next most important statistic is a measure of "spread." For example, a buyer in a crisp factory testing different packing machines would be interested to know the mean number of crisps each machine put into bags, but it is equally important to know how *consistent* the machines are.

Standard deviation: The *standard deviation* provides a measure of how much results deviate, on average, from the mean.

Although there is a formula for calculating standard deviation, you are expected to use your calculator. Make sure you understand how to enter a frequency table into your calculator and how to obtain results for the mean and standard deviation.

Try calculating the SD of weight of peanuts in these 80 packets:

Weight	No of packets
$80 \leq W < 85$	5
$85 \leq W < 90$	10
$90 \leq W < 95$	15
$95 \leq W < 100$	26
$100 \leq W < 105$	13
$105 \leq W < 110$	7
$110 \leq W < 115$	4

To save my typing fingers, I shall use SD as an abbreviation for standard deviation.

You should find that the mean weight is 96.8 and the standard deviation is 7.41.

As a rough indicator, the majority of results in a reasonably symmetrical distribution are within two standard deviations of the mean (ie $\bar{x} \pm 2SD$). For example, a class takes a mathematics test. The mean score is 65% and the standard deviation is 8%. This means that most scores will be in the range 65 ± 16, ie 49% to 81%.

The *variance* is a useful statistic for further calculations, but does not have much significance on its own. It is the square of the standard deviation.

Outliers: The "two standard deviation test" gives us a useful way of identifying outliers.

Example: The following times, in seconds, were recorded in a race:

140, 148, 152, 155, 156, 156, 157, 160, 162, 162, 165, 170.

What evidence is there to suggest that 140 s was an exceptionally fast time for this group?

Solution: Using a GDC, mean = 156.9 and SD = 7.60, so 2SD below the mean is 141.7

140 is therefore an outlier and can be classed as "exceptionally fast."

In the previous example this doesn't mean that the result can be discounted – some genuine results will be outliers. However, if the result had been recorded as 14 s this is clearly an error and should not be included in the data set.

Effect of changes to the data: Take a group of 10 children whose mean age is 12.4 years. What will be their mean age in 5 years' time? Since each of their ages will have had 5 added on, the mean will have increased by the same amount and will therefore be 17.4 years. And how will the standard deviation have changed? Not at all, since the *spread* of their ages around the mean will be exactly the same.

However, suppose a group of people take an exam marked out of 50; the mean is 35.2, and the standard deviation is 6.1. The scores are turned into percentages by doubling: the mean will now be 70.4, and the SD will have doubled as well to 12.2 since the marks will all have doubled their distance from the mean.

Thus, if a set of data has mean m and SD s, then the following rules apply:

- Add a to each of the data values: the mean will be $m + a$, and the SD will be s.
- Multiply each of the data values by b: the mean will be mb, and the SD will be sb.

> These rules can be combined. If a set of data is doubled, and then 5 added on, the mean will be $2m + 5$ and the SD will be $2s$.

Interquartile range: The standard deviation of a distribution gives us a measure of the spread of the results which is calculated using each of the values. A cruder measure of the spread is the *interquartile range* which is calculated by subtracting the lower quartile from the upper quartile. Effectively, it tells us the spread of results for the middle 50% of the population. In questions, you will normally find the standard deviation "paired" with the mean, and the IQR paired with the median.

> Not to be confused with the range which is simply the difference between the maximum and minimum values.

A survey is carried out to find the waiting time of 100 customers in a post office.

(a) Calculate an estimate of the mean waiting time.
(b) Construct a cumulative frequency table for the data.
(c) Use the table in (b) to draw a cumulative frequency graph, using a scale of 1 cm per 20 seconds on the horizontal axis and 1 cm per 10 customers on the vertical axis.
(d) Use the cumulative frequency graph to find estimates for the median and interquartile range.
(e) Use the graph to estimate how many people waited more than 115 seconds.

Waiting time (sec)	Number of customers
0–20	5
20–40	18
40–60	30
60–80	22
80–100	9
100–120	7
120–140	6
140–160	3

4. STATISTICS AND PROBABILITY

(a) Mean = 64.4 (GDC)

(b)

Waiting time (sec)	Cumulative frequency
≤ 20	5
≤ 40	23
≤ 60	53
≤ 80	75
≤ 100	84
≤ 120	91
≤ 140	97
≤ 160	100

(c)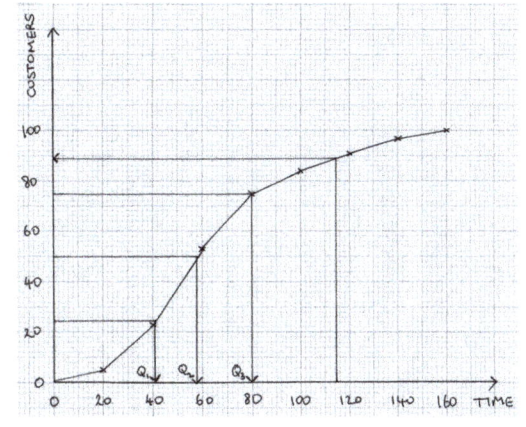

(d) Median (also known as Q_2) = 58. Lower quartile (Q_1) = 41, Upper quartile (Q_3) = 80, IQR = 80 − 41 = 39

(e) 100 − 89 = 11 people waited more than 115 minutes

You must draw the relevant lines on the graph to show how you arrive at your answers. You will be given a little leeway with the numbers, but try to be as accurate as possible.

Outliers (again): Since the IQR is a measure of spread we can also use it to define outliers. A well-used definition of an outlier is a data value which is more than 1.5 × IQR above the upper quartile, or below the lower quartile.

In the question above, Q_1 − 1.5 × IQR = 41 − 1.5 × 39 = −17.5, so it is impossible for there to be outliers at the lower end of the distribution. But Q_3 + 1.5 × IQR = 80 + 1.5 × 39 = 138.5; any values above 138.5 can be considered as outliers. We know that there must be at least 3 outliers since there are 3 values above 140 minutes.

Example: The cumulative frequency diagram shows the marks out of 70 gained by 80 students in a test. Write down the median, the quartiles and the IQR. One student took the test late and gained 66 marks – is this an outlier?

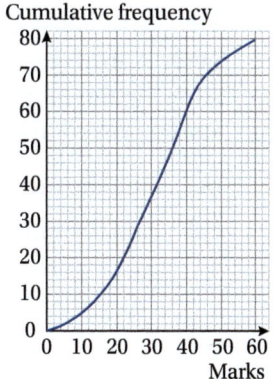

Solution: Median = 32, Q_1 = 22, Q_3 = 40, IQR = 40 − 22 = 18.

Outliers would be greater than 40 + 27 = 67. Therefore, the mark is not an outlier.

Box and Whisker plots: A box and whisker plot is a useful device for illustrating some key statistics for a distribution. The ends of the box represent the lower and upper quartiles, and the ends of the "whiskers" the extreme values. The median is shown by a line inside the box. A scale is drawn below the box and whisker plot, and different distributions can be compared.

MATHEMATICS: ANALYSIS AND APPROACHES HL

The illustration below shows the box and whisker plots for two math exams taken by a group of students.

An outlier would be indicated by a cross.

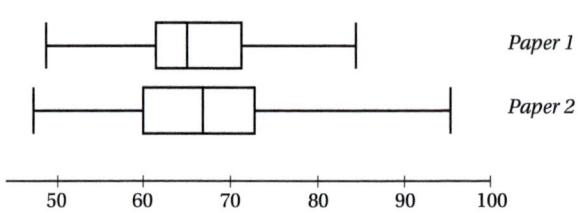

A question may ask you to compare the distributions – simple statements will suffice. For example:

- The range of results on Paper 1 is smaller.
- The median mark of the two papers is about the same.
- The interquartile range on Paper 2 is larger.

Averages and Spread: Practice Exercise

1. The following table shows the number of errors per page in a 100 page document.

Number of errors	0	1	2	3	4	5
Number of pages	28	22	18	14	12	6

a) State whether the data is discrete or continuous.
b) Draw a bar chart to represent the data.
c) Find the mean number of errors per page.
d) Find the median number of errors per page.
e) Write down the modal number of errors per page.

2. The cumulative frequency graph has been drawn from a frequency table showing the time it takes two hundred students to complete a computer game.

   (a) Find the median.
   (b) Find the interquartile range.

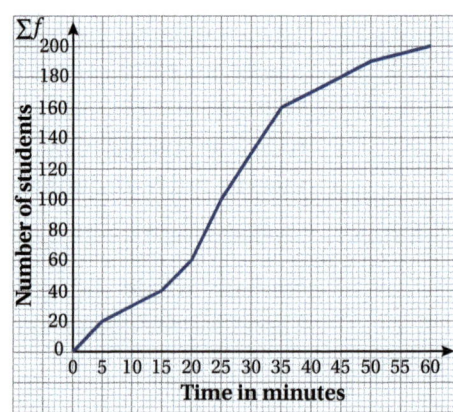

The graph has been drawn using the data in the following frequency table:

Time (min)	$0 < x \leq 5$	$5 < x \leq 15$	$15 < x \leq 20$	$20 < x \leq 25$	$25 < x \leq 35$	$35 < x \leq 50$	$50 < x \leq 60$
No. of students	20	20	a	40	60	b	10

Answers

1. (a) Discrete
   (b) [bar chart]
   (c) 1.78 (d) 1.5 (e) 0
2. (a) 25 (b) 14
   (c) $a = 20, b = 30$ (d) 25.6
3. (a) 6 (b) 6.1 (c) 1.22
4. (a) 48 (b) $a = 43, b = 80$
   (c) 20

(c) Using the graph, find the values of *a* and *b*.

(d) Calculate an estimate of the mean time taken to complete the computer game.

3. The bar chart shows the number of hours a professional musician practises each day during April:

   (a) Write down the modal number of hours.

   (b) Calculate the mean number of hours he practises each day.

   (c) Find the standard deviation.

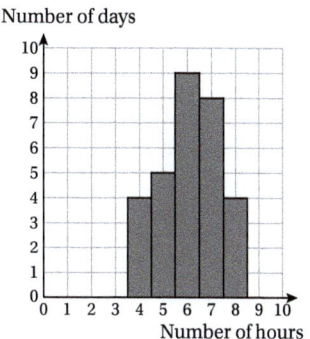

4. The following diagram is a box and whisker plot (not to scale) for a set of data. The interquartile range is 15 and the range is 50.

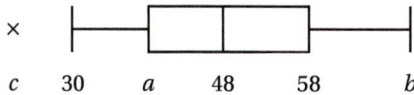

   (a) Write down the median value.

   (b) Find the values of *a* and *b*.

   (c) The cross represents an outlier. What is the highest possible integer value for *c*?

4.4 Correlation

Scatter diagrams: Two sets of data which appear to have a relationship between them are said to be *correlated*. For example, a company may find that there is a direct relationship between the amount it spends on advertising and its sales figures. Note that correlation does not imply causality: the correlation may be coincidental, or it may be linked to a third factor (perhaps, in this case, differing economic conditions).

A simple way to assess possible correlation is to draw a *scatter diagram*. The two sets of data are plotted on a standard *x*-*y* graph (but not joined in any way).

Qualitative conclusions which can be drawn about the correlation are:

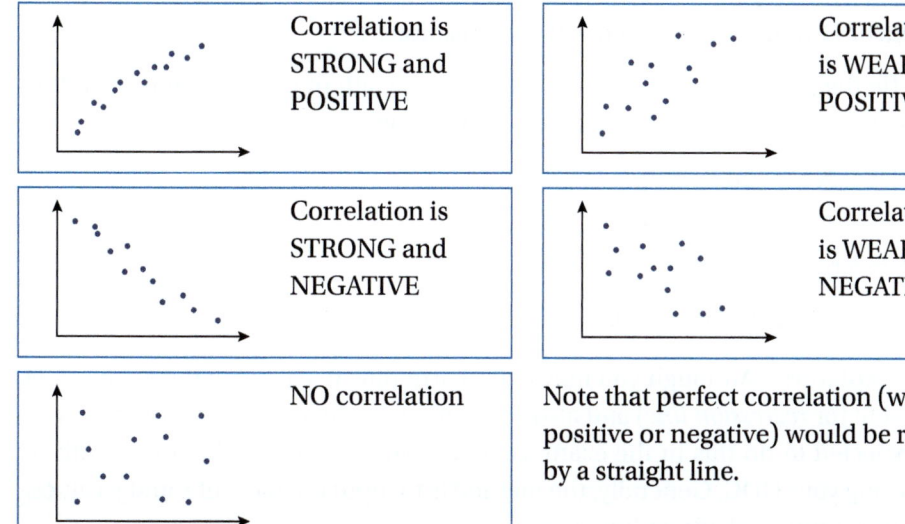

It is not necessary for the axes in a scatter diagram to be labelled from 0. We are only interested in the relationship between the points.

Note that perfect correlation (whether positive or negative) would be represented by a straight line.

Let's look at a couple of examples:

In this first one, a group of 10 to 16 year old boys were timed running 100 m. We can see that there is strong negative correlation between their ages and their times.

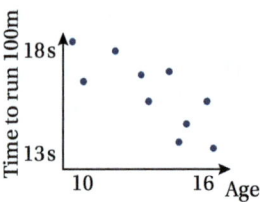

Do you think that you can extrapolate to estimate the time taken by a 21 year old? A 60 year old?

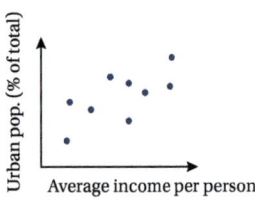

In this second graph, we are comparing countries by the percentage who live in cities against the average income per person in each country. Quite strong correlation – but, again, could we extrapolate and say that the figures will always increase together?

The answer in both cases is "no." In the first example, extrapolation would indicate that a 40 year old could run 100 m in 0 seconds! I would expect the graph to begin to turn soon, and then start back up again. In the second example, it's possible that extrapolation would work for a short distance (although we haven't been given scales); however, no country can have more than 100% of its people living in cities.

Line of best fit: A scatter diagram indicates the relationship between two variables. If we conclude that there *is* a relationship, we can draw in the "line of best fit" by eye and then use this to predict more pairs of values. If you know the mean values of the two variables, the line of best fit should pass through the point $(\bar{x}, \bar{y})$. Note that

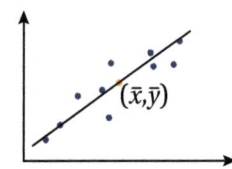

although *interpolation* (ie putting new points in between existing points) is fairly safe, *extrapolation* (ie continuing the line beyond the existing points) may not be valid. There may be reasons why the relationship does not continue in the same way.

Correlation Coefficient: For a quantitative assessment of correlation we can calculate the *product-moment coefficient*, denoted by r. This is derived from all the pairs of values and has the following properties:

- A coefficient of –1 indicates perfect negative correlation.
- A coefficient of 0 indicates no correlation.
- A coefficient of +1 indicates perfect positive correlation.

The size of r (ie the positive value of r) indicates the strength of the correlation, but this also depends on the number of pairs of values. However, we can say generally that:

- $0.25 \leq r < 0.5 \Rightarrow$ weak correlation
- $0.5 \leq r < 0.75 \Rightarrow$ moderate correlation
- $0.75 \leq r < 1 \Rightarrow$ strong correlation

(and similarly for negative values of r).

Using your calculator: Although you may have learnt how to calculate the equation of the line of best fit (or *regression line*) and also correlation coefficients using formulae, you will not be expected to do this in the exam. However, you should be able to do both of these things using your GDC. Generally, the method is to input the pairs of x and y values, then use the appropriate calculator functions.

4. STATISTICS AND PROBABILITY

Check that you are able to carry out these calculations using the following data:

x	2	4	5	7	9	10	11	15
y	3	4	6	6	7	9	10	11

You should find that the correlation coefficient is 0.97 (not surprising when you look at how closely the y values follow the x values) and that the regression line of y on x has equation $y = 1.89 + 0.65x$.

The equation of the line can be used to predict further data points. For example, what is the likely value of y when $x = 7.8$, and when $x = 17$?

When $x = 7.8$, $y = 1.89 + 0.65 \times 7.8 = 6.96$

When $x = 17$, $y = 1.89 + 0.65 \times 17 = 12.94$

However, as mentioned earlier, the latter result must be treated with caution since it has been extrapolated beyond the end of the known data – there is no guarantee that the relationship between x and y will continue to hold.

Since the equation of the regression line is a linear function, the values of a and b can represent physical quantities. a is the gradient, so represents "y quantity per x quantity". And b is the y-intercept, so represents the value when $x = 0$, which is less likely to be meaningful. Using the first example above (time to run 100 m versus age in years), a would represent "time to run 100 m per year"; in other words, how much faster a boy would run for every extra year in age. b would represent "time to run 100 m at age 0" which is clearly a nonsense; in any case, the data is only valid between ages 10 and 16.

> Because the regression line is "y on x" it can only be used to calculate y values given x values.
>
> If you are given the equation of an "x on y" regression line, that can be used to calculate x values.

The following table shows the amount of fuel (y litres) used by a car to travel certain distances (x km).

Distance (x km)	50	80	125	160	195
Fuel (y litres)	4.6	7.1	10.9	13.9	16.9

This data can be modelled by the regression line with equation $y = ax + b$.

(a) (i) Find the values of a and b.

    (ii) Explain what the gradient a represents.

(b) Use the model to estimate the amount of fuel used to drive 100 km.

(c) Could the model be used to estimate the amount of fuel to drive 250 km?

---

a) (i) $a = 0.085$, $b = 0.326$

   (ii) litres/km travelled

b) 8.82

c) Yes, because fuel usage doesn't change with distance.

   OR

   No, because this would involve extrapolating beyond the data range.

a) It's very easy to test whether your answer is correct – just try using the formula on one of the x values and check you get the corresponding y value (or something close to it).

c) Either answer would do – as long as you give a valid explanation. But it's perhaps safer to always say no to extrapolation.

MATHEMATICS: ANALYSIS AND APPROACHES HL

Piecewise models: When a function is graphed as separate line segments meeting at common points, as shown in the figure, it is known as a *piecewise linear function*. Sometimes correlation on a scatter diagram would be better achieved with more than one line of best fit, in which case we end up with a piecewise model. In other words, a single regression line may fit reasonably well, but two (or more) would fit the points much more closely.

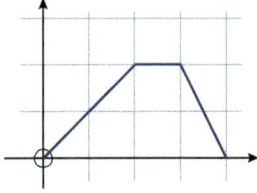

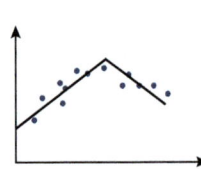

Here's an example of a piecewise model with two clearly defined sections. It's important that the point where the lines join satisfies the equations of both of the two regression lines. What sort of situations might give rise to a piecewise model? Well, the scatter diagram could represent measures of air pollution at different distances from a city centre. Moving out from the centre we pass through an increasingly industrialised zone with greater pollution; and then further out in the suburbs, so pollution decreases.

Example: A piecewise linear model contains two regression lines with equations $y = 1.6x + 2$ for $0 \leq x \leq 2$, $y = 7 - 0.9x$ for $2 \leq x \leq 5$. Find the point where the two lines meet, and also calculate y values when $x = 0.8$ and when $x = 3.5$.

Solution: Solving the two equations simultaneously (GDC) we find the lines meet at (2, 5.2). When $x = 0.8$ (first section), $y = 4.28$. When $x = 3.5$ (second section), $y = 3.85$.

> Or you can tell from the domains in the question that the two lines meet when $x = 2$.

4.5 Sampling Methods

Ideally, statistics are gathered from a whole population. In practice, this is usually too expensive and too time-consuming, so a population sample is used instead. The difficulty is to make the sample representative of the population, so that conclusions drawn from the sample can be applied to the population.

"Population" doesn't just refer to people. For example, a machine in a factory may turn out 1000 components in a day. From this population, the quality control manager may want to select a sample of 20 components to test – how should he set about it?

One of the aims of sampling is to introduce as little bias as possible.

- Standing on a street corner at 11am selecting people for a survey introduces bias because the majority of the sample will not include people who are employed.
- Asking people to review a restaurant online with the promise of a possible discount introduces bias because people are more likely to give a good review.
- Selecting all 20 components (for the quality control test) in the morning may introduce bias because the machine could become more erratic later in the day.

Simple random sampling: The definition of a simple random sample is that every member of the population is equally likely to be chosen for the sample. Suppose you want a sample of 25 employees to be chosen from a company employing 300. Write all 300 names on pieces of paper, put them in a box, and then pull out 25 names. Every employee is thus equally likely to be in the sample. What this method will not achieve is a sample which reflects the make-up of the population: the male/female split, the number of employees in each age group and so on. By chance, the sample might contain all women aged between 25 and 30.

> Instead of pieces of paper, every employee could be allocated a number, and then the sample is selected using a computerised random number generator.

4. STATISTICS AND PROBABILITY

Convenience sampling: In this method of sampling, data is collected from population members who are conveniently available. In other words, no inclusion criteria are specified before sampling takes place. Its main purpose is to gain some initial data prior to a proper study taking place – for example, to obtain a perception of an image brand by simply going up to people in the street and asking their opinion.

Systematic sampling: Returning to our factory producing 1000 components a day, a simple way of selecting a sample of 20 is to choose every 50th component off the production line. If the first sample is chosen randomly from the first 50 components, then this ensures every component has an equal chance of being chosen. The main disadvantage of this method is that we need to know the population size to begin with. If, for example, a researcher wants to study a sample of 20 trees in a forest, she cannot systematically choose them without knowing how many trees there are in total.

Quota sampling: Quota sampling is used to select members of a population which has been divided into sub-groups. For example, a researcher wants to gather data on males and females, further sub-divided into age under 21 or 21 and over. He requires 30 of each (ie a sample size of 120), and simply approaches people until he has fulfilled each quota. This method is used if time or funding is limited but, since the sample is not genuinely random, the data is unreliable. Further bias could be introduced because the researcher may only question people who look approachable.

Stratified sampling: This method is the same as quota sampling except that within each group the sample is chosen by a method such as simple random sampling. For example, a school has 460 boys and 540 girls. A representative sample of 50 is to be chosen using stratified sampling. 50 is $\frac{1}{20}$ of the whole school, so we need to choose $\frac{1}{20} \times 460 = 23$ boys and $\frac{1}{20} \times 540 = 27$ girls. Now put all the boys' names in a box and choose 23 of them, then do the same for the 27 girls. Alternatively, go down the school list choosing every 20th boy and every 20th girl.

Stratified sampling will be used, for example, when it is important to an opinion pollster to gather results according to gender, age, political persuasion, and employment type.

MATHEMATICS: ANALYSIS AND APPROACHES HL

The top two year groups in a school (Year 12 and Year 13) are to take part in a survey about the future of school uniform. The 200 pupils are divided by gender and year group as follows:

	Male	Female
Year 12	55	50
Year 13	48	47

A sample group of 30 are to be chosen. How would you select the group using stratified sampling?

There are 200 pupils in the group. $30 \div 200 = 0.15$.

Year 12 male $= 0.15 \times 55 = 8.25$ Select 8.

Year 12 female $= 0.15 \times 50 = 7.5$ Select 8.

Year 13 male $= 0.15 \times 48 = 7.2$ Select 7.

Year 13 female $= 0.15 \times 47 = 7.05$ Select 7.

Now allocate numbers to each of the members of the group, for example Year 12 males from 01 to 55, and use a random number generator to select the appropriate number from each group.

If the numbers don't come out as integers, it is important to ensure that the sample size from each group still adds up to the overall sample size. In this case I rounded 7.5 up to 8 to ensure the total was 30.

Any method of random sampling can be used for the second part.

4.6 Probability Notation and Formulae

Notation: The *sample space* in a given situation is the set of all the things that can happen and is defined by the letter U. An *event* is one of the things that can happen and is given any other capital letter. A capital P stands for "probability", so we can shorten "the probability of event A" to $P(A)$. The number of ways A can happen is denoted by $n(A)$. Probabilities are always numbers between 0 (definitely won't happen) and 1 (definitely will happen). The probability that A happens is given by $P(A) = \dfrac{n(A)}{n(U)}$. The probability that event A does ***not*** happen is denoted by A'. It follows that $P(A) + P(A') = 1$.

The set notation symbols $\cap$ and $\cup$ are used for the words "and" and "or" in probability.

Combined events: The probability of event A ***or*** event B happening (and this includes both) is calculated using addition.

- $P(A \cup B) = P(A) + P(B)$

but this formula works ***only*** if A and B are *mutually exclusive* – ie they cannot both happen. If they are not mutually exclusive, use:

- $P(A \cup B) = P(A) + P(B) - P(A \cap B)$

The probability of events A and B ***both*** happening is calculated by multiplication (remember that multiplying fractions gives a ***smaller*** answer and it is ***less*** likely that both events will happen than just one).

4. STATISTICS AND PROBABILITY

- $P(A \cap B) = P(A) \times P(B)$

but this formula works *only* if A and B are *independent* – ie one of them happening does not affect the probability of the other happening. If the events are not independent we are into the realms of *conditional probability* – ie the probability of one event happening if another has already happened. This is written as $P(A|B)$, and read as "the probability of A given B."

- $P(A|B) = \dfrac{P(A \cap B)}{P(B)}$

> A bag contains balls of two different colours. One is taken out, then another. The colour of the second is independent of the first if the first has been put back. If the first has been kept out, the colour of the second **depends** on the colour of the first.

Note that the definition of independence is $P(A) = P(A|B) = P(A|B')$ (in other words, the probability of A is the same whether or not B has happened). But if you are asked to test whether events are independent, the normal test is to check if $P(A \cap B) = P(A) \times P(B)$.

For the events A and B, $P(A) = 0.3$, $P(B) = 0.4$.

(a) Find $P(A \cup B)$ if A and B are independent events.

(b) Find $P(A' \cap B')$ if A and B are mutually exclusive events.

(a) $P(A \cap B) = 0.3 \times 0.4 = 0.12$ $P(A \cup B) = P(A) + P(B) - P(A \cap B)$ $= 0.3 + 0.4 - 0.12$ $= 0.58$	*(a) We are not told the events are mutually exclusive so we must use the full formula or P(A or B). This involves P(A and B) which we can calculate because they are independent.*
(b) $P(A \cup B) = P(A) + P(B) = 0.7$ So, $P(A' \cap B') = 1 - P(A \cup B) = 0.3$	*(b) is a new question so we cannot use independence. You will see in the next section how a Venn diagram can help you solve these sorts of problems more easily.*

The formulae can be quite difficult to use, so only use them if you **have** to. Many probability questions can be solved by using appropriate diagrams as shown on the next few pages.

4.7 Lists and Tables of Outcomes

Lists: A list of possible outcomes is useful if there aren't too many of them; and it is important to ensure that each outcome in the list is equally likely. For example, when three coins are thrown, the possible combinations of heads and tails are:

HHH, HHT, HTH, HTT, THH, THT, TTH, TTT

If we want to find P(exactly two heads) we can see that there are three ways of achieving this (HHT, HTH, THH) so the probability is 3/8.

Possibility Space diagram: This is a way of showing a list of outcomes on a diagram, but can only be used for two events. For example, the following diagram shows all the possible totals when two six-sided dice (red and green) are thrown.

MATHEMATICS: ANALYSIS AND APPROACHES HL

Note that there is only one way a double 2, say, can happen – a 2 on the green and a 2 on the red. But a 1 and a 3 can happen in two ways: 1 on the green and 3 on the red, or the other way around.

Green							
6	7	8	9	10	11	12
5	6	7	8	9	10	11
4	5	6	7	8	9	10
3	4	5	6	7	8	9
2	3	4	5	6	7	8
1	2	3	4	5	6	7
	1	2	3	4	5	6

Red

Thus there are 36 possibilities. Some examples of probabilities are:

$P(\text{Total of } 5) = 4/36$

$P(\text{Total of 5 or 7}) = 10/36$

$P(\text{Total of 4 or a double}) = 8/36$

$P(\text{Double}|\text{total} \geq 9) = 2/10$

The conditional probability in the last example is easy to see on the diagram. We *know* that the total is ≥ 9, and there are 10 ways this can have happened. Of these, 2 could be a double.

Tables of outcomes: Tables of outcomes show how many ways two events can, or cannot, happen. For example, let's take a survey of 200 people of whom 90 are female. 60 people were unemployed, including 20 males. Filling that information into a table of outcomes, we get:

	Males	Females	Totals
Unemployed	20		60
Employed			
Totals		90	200

You will see that there is just enough information to allow us to fill in the rest of the table. Try it before looking at the answer below.

	Males	Females	Totals
Unemployed	20	40	60
Employed	90	50	140
Totals	110	90	200

Now, if a person is selected at random from the 200, what is the probability that the person is (a) an unemployed female, (b) a male, given that the person is employed.

(a) There are 40 unemployed females out of 200, so

$$P(\text{unemployed female}) = \frac{40}{200}.$$

(b) Knowing that the person is employed, he/she must be one of the 140. Of these 90 are males, so $P(\text{male}|\text{employed}) = \frac{90}{140}$. As with the possibility space diagram, it is easy to deal with conditional probability when using a table of outcomes.

4.8 Venn Diagrams

In a room there are 20 people. 11 have black hair, 6 have glasses. 2 people have both black hair and glasses. Imagine that we draw two circles on the floor labelled "black hair" and "glasses" and ask the people to stand in the appropriate circle. The circles will have to overlap to allow for the two people with both. The numbers of people in each region of the room will be:

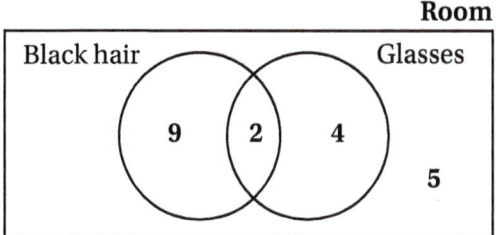

This is the same as a Venn Diagram. The "room" represents the sample space – for a particular question, there is nothing outside. Each circle represents a set, the overlap is the intersection.

Some examples of Venn Diagrams are shown below:

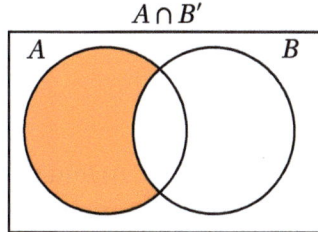 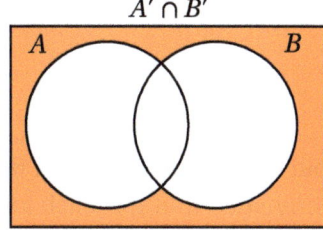 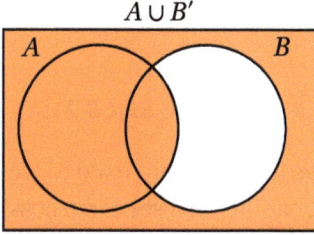

Points to note when filling in the numbers in a Venn Diagram:

- Start at the centre. If you are not told how many in the intersection, work it out like this: suppose you know there are 15 people in total in the two circles, 10 in circle A and 8 in circle B. 10 + 8 = 18, 3 more than 15, so there are 3 in the intersection.
- When we were told that there were 11 people with black hair, this *includes* those with both black hair and glasses. Same with the 6 people with glasses.
- Don't forget to fill in the outer region – although in some questions this set will be "empty."

Probabilities can now be calculated easily. When someone is selected at random, the probability they have:

Black hair and glasses = 2/20

Black hair and no glasses = 9/20

Not got glasses = 14/20

Glasses or black hair (or both) = 15/20

Glasses given black hair = 2/11

Glasses given not black hair = 4/9

Venn Diagrams are very helpful when calculating conditional probability – you may like to look at an article I have written for a fuller explanation.

Additional resources are on the website www.peakib.com.

The next example shows how a Venn diagram can be used as an alternative to using the formulae.

Example: A and B are independent events. $P(A \cap B) = 0.2$, $P(A \cap B') = 0.3$

Find $P(A \cup B)$.

Solution: Intersections are easy to draw on a Venn diagram – see right.

Now we note that A and B are independent so we can use the formula

$P(A \cap B) = P(A) \times P(B)$.
$0.2 = 0.5 \times P(B) \Rightarrow P(B) = 0.4$

Using this information we can complete the Venn diagram, and hence answer the question: $P(A \cup B) = 0.7$.

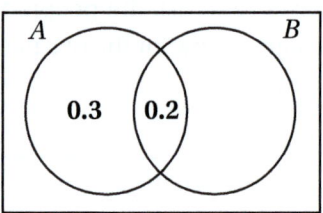

We can answer most questions once we have a complete Venn diagram. For example, can you show that $P(A|B) = 0.5$ and also that $P(B'|A) = 0.6$

4.9 Tree Diagrams

Tree diagrams are used to work out the probabilities for a *succession* of events. To find the probability of a set of successive branches, multiply each individual probability *along* the branches. To find the probability of one of several branches occurring, add the probabilities of each outcome.

Note that the probabilities associated with, say, taking two balls out of a bag simultaneously are the same as if the balls were taken out consecutively.

eg: P(rains today) = 0.3. If it rains today, P(rains tomorrow) = 0.65

However, if it is dry today, P(rains tomorrow) = 0.2 The tree diagram which shows the full set of possible outcomes and their associated probabilities is:

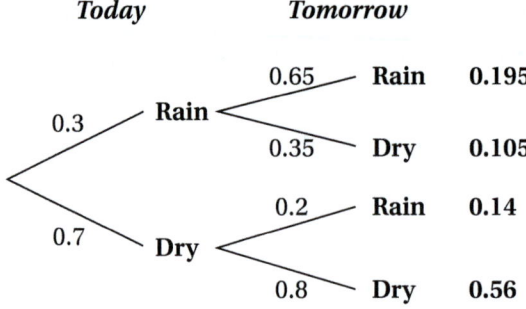

Note the following points:

- Probabilities of branches coming out of one point add to give 1 since they cover all possibilities.
- The overall probabilities also add to give 1.
- The weather tomorrow is *not* independent of the weather today, hence the different probabilities depending on today's weather.

Some example probabilities are:

- P(two rainy days) = 0.195
- P(at least one rainy day) = 0.195 + 0.105 + 0.14 = 0.44
  $\qquad\qquad\qquad\qquad$ = 1 − P(two dry days)
- P(exactly one rainy day) = 0.105 + 0.14 = 0.245

4. STATISTICS AND PROBABILITY

Questions about tree diagrams often come with a sting in the tail in the form of a conditional probability problem. Consider the following tree diagram where event A = "my alarm clock works" and event L = "I am late for school."

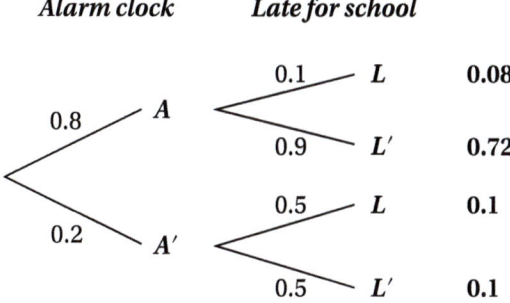

P(I am late for school) = $P(L|A) + P(L|A') = 0.08 + 0.1 = 0.18$.

But suppose I am late for school, and my teacher says: "Obviously your alarm clock didn't work." What is the probability she is right?

The calculation is $P(A'|L) = \dfrac{P(A' \cap L)}{P(L)} = \dfrac{0.1}{0.18} = 0.556$. Another way to think of this is to consider the expected values over 100 days. On 18 of them we would expect to be late, and of these 18 we would expect the alarm clock to have failed on 10.

> The probability that Hector studies for an exam is 0.7. If he studies, the probability that he passes the exam is 0.8. If he does not study, the probability that he passes is 0.5.
>
> (a) Draw a tree diagram to represent the information.
> (b) Find the probability that he passes the exam.
> (c) Given Hector passes the exam, find the probability that he studied.
>
> (b) 0.71 (c) 0.789

See worked solution online

Expected number of occurrences: In the previous example about alarm clocks, I suggested looking at the number of times I would expect to be late over a period of 100 days. P(late) = 0.18, so I would expect to be late on $0.18 \times 100 = 18$ days. This is a specific example of a simple calculation: whenever you have the probability p of an event, and that event could happen on n occasions, the expected number of occurrences is pn.

Bayes' Theorem: The two questions are both examples of *Bayes' Theorem*. The formula can be derived from generalising the tree diagram…

```
              P(B|A)      B      P(A ∩ B)  = P(A) × P(B|A)
        P(A) /  A
            <
              P(B'|A)     B'     P(A ∩ B') = P(A) × P(B'|A)

              P(B|A')     B      P(A' ∩ B) = P(A') × P(B|A')
       P(A') \ A'
            <
              P(B'|A')    B'     P(A' ∩ B')= P(A') × P(B'|A')
```

… and we can see that:

$$P(A|B) = \dfrac{P(A \cap B)}{P(B)} = \dfrac{P(A) \times P(B|A)}{P(A) \times P(B|A) + P(A') \times P(B|A')}$$

117

However, it is generally easier to obtain solutions from the tree diagram than from the formula. You are expected to use Bayes' Theorem for up to three events – try the following example.

Example: A computer manufacturer receives a particular component from three different suppliers. Company A supplies 40% of the total, company B supplies 25% of the total, and company C supplies the rest. 4% of the components supplied by A are defective; 3% from B are defective; and 5% from C are defective.

(a) 100 components are checked at random. How many would be expected to be defective.

(b) What is the probability that a component which is found to be defective has been supplied by company C.

Solution: (a) P(defective) = 0.41, so 4 defectives would be expected

(b) P(C|defective) = $\frac{0.0175}{0.041}$ = 0.427

4.10 Discrete Probability Distributions

A probability distribution shows the probabilities for all the outcomes of a particular event. Discrete probability distributions relate to events which can only have certain outcomes – usually in the form of integers.

Uniform distributions: If all the outcomes are equally likely, the distribution is called *uniform*. For example, here is the probability distribution for the random variable X where X represents the outcomes when throwing a die.

> Note that the capital letter X is used to describe the random variable, whereas lower case x is used to represent the actual values.

x	1	2	3	4	5	6
P($X = x$)	1/6	1/6	1/6	1/6	1/6	1/6

Distributions defined by a function: The following is an example of a probability distribution defined by a function:

$$P(X = x) = \begin{cases} kx, & x = 1, 2, 3, 4, 5 \\ 0 & \text{otherwise} \end{cases}$$

This means that x can only take values 1 to 5, and has probability kx for these values. The best thing to do is put all the information into a table:

x	1	2	3	4	5
P($X = x$)	k	$2k$	$3k$	$4k$	$5k$

In all probability distributions, the probabilities add to give 1, so $15k = 1$, giving $k = \frac{1}{15}$.

We can fill the probabilities into the table:

x	1	2	3	4	5
P($X = x$)	$\frac{1}{15}$	$\frac{2}{15}$	$\frac{3}{15}$	$\frac{4}{15}$	$\frac{5}{15}$

4. STATISTICS AND PROBABILITY

Expected value (mean): By multiplying each value of x by its associated probability, we obtain the *expected mean*. Thus the formula is: $E(X) = \sum xp$. In the above example we get $\frac{55}{15} = 3.67$, and the more times we carry out the trial, the closer the ***actual*** mean will get to this value.

Example: The probability distribution for a random variable X is given by:

$P(X = x) = kx(x - 1)$, for $x = 2, 3, 4, 5, 6$

(a) Find the value of k; (b) Find the expected mean of the distribution.

Solution: First we must draw up the probability table:

x	2	3	4	5	6
$P(X=x)$	$2k$	$6k$	$12k$	$20k$	$30k$

Thus $70k = 1 \Rightarrow k = \frac{1}{70}$.

Now we can fill the probabilities into the table and work out the expected mean.

x	2	3	4	5	6
$P(X=x)$	$\frac{2}{70}$	$\frac{6}{70}$	$\frac{12}{70}$	$\frac{20}{70}$	$\frac{30}{70}$

$E(X) = 2 \times \frac{2}{70} + 3 \times \frac{6}{70} + 4 \times \frac{12}{70} + 5 \times \frac{20}{70} + 6 \times \frac{30}{70} = \frac{350}{70} = 5$

Expected variance: The formula for the expected variance of a discrete probability distribution is $\text{Var}(X) = \sum x^2 p - [E(X)]^2$. This is often remembered as "the expectation of the squares minus the square of the expectation." The expected standard deviation will, of course, be the square root of the expected variance.

The following table shows the probability distribution of a discrete random variable X:

x	-1	0	1	2	3
$P(X=x)$	0.2	$10k^2$	0	0.4	$3k$

(a) Find the value of k.
(b) Find the expected value of X.
(c) Find the expected standard deviation of X.

(a) $10k^2 + 3k + 0.6 = 1 \Rightarrow k = 0.1$ (GDC)
(b) $E(X) = -0.2 + 0 + 0.8 + 0.9 = 1.5$
(c) $\text{Var}(X) = (-1)^2 \times 0.2 + 2^2 \times 0.4 + 3^2 \times 0.3 - 1.5^2$
    $= 2.25$
Standard deviation $= \sqrt{2.25} = 1.5$

The quadratic could have been solved without a GDC. Multiply through by 10 (to make all the coefficients integers), then divide by 2 to simplify. Then factorise.

Games of chance: Let's play a game. You throw two dice. If you get a 9 or 11, I'll give you $3; if you get a double, I'll give you $2. The catch is, you must pay me $1 to play. Is it worth it? We can draw up a table of probabilities (see page 113 for how to deal with totals of two dice).

Event	Prob.	Outcome
9 or 11	6/36	$3
Double	6/36	$2
Other	24/36	$0

The expected mean is $\frac{6}{36} \times 3 + \frac{6}{36} \times 2 + \frac{24}{36} \times 0 = \frac{30}{36}$. Thus, on average, you can expect to win under $1 per game, so you will lose out in the long run – and you will decline my offer to play. (Moral: you can't make money out of IB Mathematics students!) You could alternatively include the $1 in the table by making the outcomes $2, $1 and –$1. This would make the expected mean $-\frac{6}{36}$.

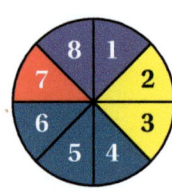

In a *fair* game, neither side has the advantage, in which case E(X) = 0. Have a look at the rather colourful spinner. Suppose we play 3 tokens for a spin, and then win as follows: Purple = 3 tokens, Yellow = 4 tokens, Blue = 1 token. What should the payout be for red to make the game fair?

Suppose X represents the distribution of net winnings (ie taking the stake of 3 tokens into account), and let p be the payout for a red; $E(X) = 0 \times \frac{2}{8} + 1 \times \frac{2}{8} + (-2) \times \frac{3}{8} + (p-3) \times \frac{1}{8}$. For a fair game, E(X) = 0 and this gives $p = 7$.

Linear transformations of X: In the same way that the mean and standard deviation of a frequency distribution are changed when the data is changed, so too are the expected mean and standard deviation of a probability distribution. If a constant is added to all the values of X then the same constant is added to the mean; and if each value of X is multiplied by a constant, then both the expected mean and expected standard deviation are multiplied by the same constant. In summary:

- $E(aX + b) = aE(x) + b$
- $\text{Var}(aX + b) = a^2 \text{Var}(X)$
- SD will be multiplied by a

4.11 The Binomial Distribution

The binomial probability distribution is a special case of a discrete distribution. You can use it when:

- there are a fixed number of "trials";
- each trial has only two possible outcomes, "success" and "failure";
- the results of each trial are independent of each other;
- the probability of success remains the same.

For example, my young child wakes me up 1 night in 4. I want to find the probability that I will be woken up 3 nights out of 10.

- The number of trials, n, is 10.
- The probability of "success" (ie being woken up!) is 0.25
- We therefore say that the distribution is $X \sim B(10, 0.25)$

📱 You will be expected to use a calculator for binomial probabilities, and understand the difference between probability and cumulative probability.

4. STATISTICS AND PROBABILITY

The calculation has three parts to it:

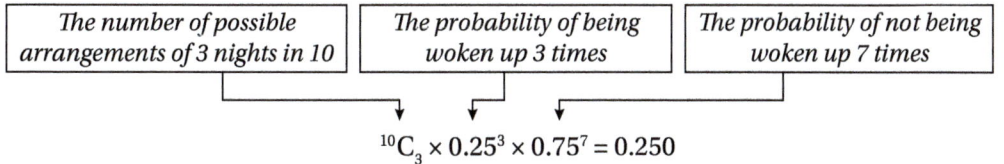

Thus there are always three parts to a binomial probability calculation *except* when you are at either end of the distribution. In which case: P(woken up all 10 nights) = 0.25^{10}; and the probability of not being woken up at all in ten nights is 0.75^{10}.

Actually, $^{10}C_{10} \times 0.25^{10} \times 0.75^0$

Getting the probability of success: You may simply be given the probability of success, or:

- You calculate the probability from previous experience (as in the example above)
- You calculate it from your knowledge of the situation (eg: success is getting a 2 on the spinner: $P = \frac{1}{3}$.
- The probability is the result of a calculation from a previous part of the question.

Example: 350 of the 500 pupils in a school have the letter "s" in their name. Six sports prizes are awarded at the end of term. What is the probability four of the prizewinners have an "s" in their name.

This question assumes that a prizewinner can win more than one prize. Why?

Solution: The first sentence gives us the probability of success.

P("s") = $\frac{350}{500}$ = 0.7. Therefore $X \sim B(6, 0.7)$.

P(4 successes) = 0.324

More than one outcome: Since binomial probabilities are all mutually exclusive (I cannot be woken up both 3 nights *and* 4 nights in 10), the probability of one of several outcomes occurring can be found by addition. Thus, P(I am woken up 3 or 4 nights out of 10) is:

$^{10}C_3 \times 0.25^3 \times 0.75^7 + {}^{10}C_4 \times 0.25^4 \times 0.75^6 = 0.396$.

Cumulative probabilities: What is the probability of being woken up on fewer than 3 nights out of 10: that is, P(0, 1, or 2). You can add these three probabilities together or use the cumulative probability function on your calculator which gives 0.526. This enables us to answer questions such as: "Find the probability that I am awoken on at least 3 nights out of 10."

Check the wording of questions carefully. It might say: "Find the probability that I have at least eight nights when I am *not* woken up." Check this also gives 0.526.

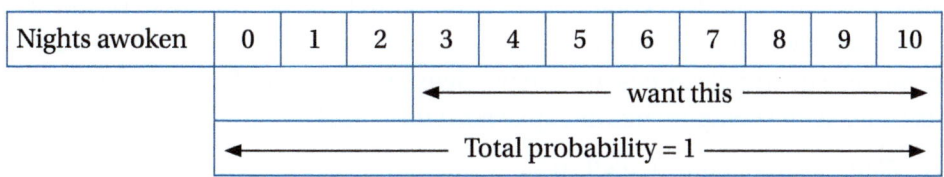

The diagram shows that the easiest way to calculate this is to find the cumulative probability up to 2, and subtract the answer from 1. This gives 1 − 0.526 = 0.474.

On some GDCs this cumulative probability could be calculated directly.

Because there is a formula for binomial distribution probabilities, examiners sometimes like to set questions which are more algebraic in nature, such as this one:

> The probability of an event occurring is p.
>
> (a) Write down an expression in terms of p for the probability that the event occurs exactly 5 times out of 8.
>
> (b) Hence find the possible values of p such that the probability of an event occurring exactly 5 times out of 8 is 0.23.

(a) $^8C_5 p^5(1-p)^3 = 56p^5(1-p)^3$ (b) $56p^5(1-p)^3 = 0.23$ $p = 0.513$ or 0.728 (GDC)	(a) We have to work out for ourselves that this involves a binomial probability. This question also illustrates that, although you will often be able to work out binomial probabilities on your calculator, you do need to know the formula as well. (b) You might like to see my article on using the TI-84 to solve equations – see the website.

> Joe is a football player. When shooting penalties, he succeeds 3 times out of every 5. In practice, he shoots 8 times. Find, to 3 significant figures, the probabilities of:
>
> (a) Scoring all 8 penalties.
> (b) Scoring 6 penalties out of 8.
> (c) Scoring at least 6 penalties out of 8.

$X \sim B(8, 0.6)$ (a) $P(X = 8) = 0.6^8 = 0.0168$ (b) $P(X = 6) = 0.209$ (GDC) (c) $P(X \geq 6) = 1 - P(X \leq 5)$ $\qquad = 0.315$ (GDC)	It's always worth stating the distribution you are going to use – the examiner can see what you're doing, and it helps sort things out in your mind too.

I've included the next question here although it could well be part of a section B question because it illustrates how to deal with wording which can send your mind in a spin. The key thing is to strip the wording down until it becomes possible to see the binomial probabilities required.

> A machine contains a critical component. This component is replicated 10 times within the machine, and the machine works as long as at least one of the ten components is working. Each has an independent probability of failing within one year of 0.7, and all the components are replaced at the end of a year.
>
> (a) Find the probability that all 10 fail within the year.
> (b) Find the probability that the machine is in operation at the end of the year.
> (c) (i) Suppose we put in n components. What is the probability that the machine is operating at the end of the year?
>   (ii) Hence find the smallest number of components to install which will ensure a probability of at least 0.99 that the machine is working at the end of the year.

4. STATISTICS AND PROBABILITY

> (a) $X \sim B(10, 0.7)$ where X = component fails
>
> $P(X = 10) = 0.7^{10} = 0.0282$
>
> (b) $P(\text{machine works}) = 1 - P(X = 10) = 0.9718$
>
> (c) (i) $X \sim B(n, 0.7)$
>
> $P(\text{machine works}) = 1 - 0.7^n$
>
>   (ii) $1 - 0.7^n > 0.99$
>
> $0.7^n < 0.01$
>
> $n > \dfrac{\log 0.01}{\log 0.7} = 12.9$
>
>   So the smallest value of n is 13

(a) The only issue is a semantic one – the failure of a component is a probability success!

(b) At least 1 component working is the same as $1 - P$(none are working); a familiar calculation in binomial probability questions.

(c) Restating the distribution parameters helps to see the connection between part (c) and part (b). Note the reversal of the inequality during the working because $\log 0.7$ is a negative number. Or solve directly on the GDC.

Why might a binomial probability not be appropriate? Refer to the conditions on page 120 under which a binomial distribution is valid. Firstly, the events must be *independent*. The answer to the question in the notes box is 0.401. However, the assumption of independence may be wrong: the 1 in 20 is an average figure over a period of time, but perhaps if the temperature in the factory rises too much, more faulty boards are produced. Then our batch of 10, if they were all manufactured together, may have a higher incidence of faults.

> It is known that 1 out of 20 printed circuit boards supplied by a certain manufacturer has a fault. What is the probability that at least 1 in a batch of 10 is faulty?

Secondly, you cannot use the binomial distribution if the probabilities change. For example, there are 10 pieces of paper folded up in a box, and three have crosses marked on them. To find the probability that, when two pieces of paper are drawn out, neither has a cross, you need a tree diagram. The probabilities change each time a piece of paper is removed.

Expected mean: Fortunately, we do not have to go through the normal process for discrete distributions – there is a simple formula for the expected mean of a binomial distribution. Suppose Joe (who appeared a few questions back) decides to enter a marathon penalty shooting competition and goes for 400 shots. How many times would he succeed? His probability of success is 0.6, so we would expect him to succeed $0.6 \times 400 = 240$ times. Thus, if $X \sim B(n, p)$, then $E(X) = np$.

> In reality, as he tires, his probability of success would probably decrease.

Expected variance: It turns out that, not only is there a simple formula for the expected mean of a binomial distribution, there is also an equally simple one for the expected variance.

> Remember that variance is the square of standard deviation.

If $X \sim B(n, p)$, then $\text{Var}(X) = np(1 - p)$. Returning once again to Joe, the expected variance of his 400 penalty shots will be $400 \times 0.6 \times 0.4 = 96$. Thus the expected standard deviation will be $\sqrt{96} = 9.80$. What does this tell us? You should recall that we generally expect results to be within two standard deviations of the mean. In this case, with a mean of 240, this gives a likely range of about 221 to 259. If Joe scored, say, 270 penalties out of 400, we might need to question the accuracy of the 3 out of 5 figure quoted in the original question. He could be better than we thought!

In the next example we see again how a binomial distribution question leads to a bit of algebra.

Example: A binomial $X \sim (n, p)$ distribution has mean 4 and variance 2.4. Find the values of n and p.

Solution: When you have to find two unknowns, the chances are you will end up with simultaneous equations. These will arise from the formulae for the mean and the variance.

$$np = 4,$$
$$np(1-p) = 2.4$$

Substitute the value of np from the first equation into the second:

$$4(1-p) = 2.4 \Rightarrow p = 0.4$$

Substitute this value of p into the first equation:

$$n = \frac{4}{0.4} = 10$$

Binomial Distribution: Practice Exercise

Answers
1. 0.0115, 1.75, 1.3125
2. 0.283, 0.795, 0.0194
3. 0.13, 0.12, so (a).
4. 17.9 to 32.1, so 18 to 32
5. $P(X > 1) = 0.081$. In a year, this happens $52 \times 0.081 = 4.2$. So about 4 times a year.
6. $n = 18$, $p = \frac{1}{3}$

1. $X \sim B(7, 0.25)$. Find $P(X = 5)$, the mean and variance of X.

2. $X \sim B(12, 0.8)$. Find $P(X = 10)$, $P(X > 8)$ and $P(X$ is less than $7)$.

3. Which is more likely: (a) $X \sim B(10, 0.15)$, $P(X = 3)$
   (b) $X \sim B(12, 0.12)$, $P(X = 3)$?

4. A coin is thrown 50 times. X represents the number of heads. What would be a likely range of values for X?

5. I work 5 days a week, and I'm late home from work about once every ten days. My partner gets cross if I'm late home more than once in a week, and I then buy them a present. How many weeks in a year would I expect to have to buy a present?

6. The random variable $X \sim B(n, p)$ has mean 4 and standard deviation 2. Determine the values of n and p.

4.12 The Normal Distribution

The Normal Distribution is used to model many commonly occurring frequency distributions, eg: the heights of trees, weights of people. The curve has the following properties:

- It is symmetrical about the mean value, μ.
- The median is the same as the mean.
- The curve approaches the x-axis asymptotically (although this is not true for the majority of distributions the curve is modelling).

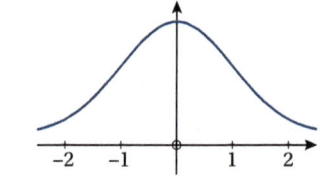

The curve (shown in the diagram) is called the *standard* normal distribution: its mean is 0, its standard deviation is 1 and the area under the curve is 1.

Standardised value: The basis of all normal distribution calculations is the *standardised value* which is the number of standard deviations that the actual value lies above or below the mean. Thus, if a group of people have a mean height 170 cm with standard deviation 10 cm, and a mean weight of 65 kg with standard deviation 5 kg, then the probability that a person chosen at random is less than 180 cm high is exactly the same as the probability of weighing less than 70 kg; both values are one standard deviation above the mean.

4. STATISTICS AND PROBABILITY

For a normal distribution with mean μ and standard deviation σ the formula for the standardised value is:

$$Z = \frac{x - \mu}{\sigma}$$

Simple normal probability calculations: Basic probabilities can be calculated on your GDC by entering four values: lower bound, upper bound, mean, standard deviation. In questions where there is no lower or upper bound, you can either use values such as -1×10^{99} and 1×10^{99}, or just any values much smaller or larger than those in the question.

When finding answers on your GDC you need to show some working. I suggest a shaded diagram will demonstrate to the examiner that you understand what is happening.

Example: A group of people are asked to carry out a simple task. The length of time taken, in minutes, follows a normal distribution where the mean is 3.2 and standard deviation is 0.6.

   (a) If a member of the group is chosen at random, find:

      (i) P(she takes between 2.5 and 3 minutes)

      (ii) P(she takes more than 3 minutes)

Solution: (a) (i) On your GDC, enter the values for the lower and upper bounds, the mean and the SD.

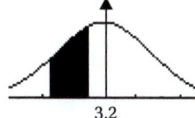

$$P(2.5 \leq X \leq 3) = 0.248$$

      (ii) Now enter 3 for the lower bound, and a number such as 100 for the upper bound.

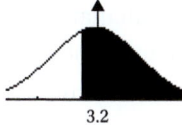

$$P(X > 3) = 0.631$$

You can also use the inverse normal function on the GDC to reverse the process. The inverse will give you the Z value associated with a particular probability, and hence the X value.

Carrying on with the previous question:

   (b) The fastest 10% of the group are then selected to perform a more advanced task. What was the slowest time required to be selected?

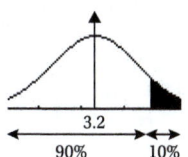

Solution: The inverse function can be used only from the left hand end of the distribution. Thus, to find the lowest value for the top 10%, we must find the highest value of the bottom 90%. Using the GDC directly we find that the slowest time was 3.97 minutes.

However, it may be advisable to show some working. Use the inverse function with $\mu = 0$, $\sigma = 1$ and area = 0.9 to give $Z = 1.28$. Then:

$$Z = \frac{X - \mu}{\sigma}, \text{ so } 1.28 = \frac{X - 3.2}{0.6}$$

Thus $X = 3.2 + 1.28 \times 0.6 = 3.97$ minutes.

📱 Use your GDC to show that $P(-1 \leq Z \leq 1)$ is about 68%. In other words, about two thirds of all values are within one standard deviation of the mean.

For example, if $\mu = 35$ and $\sigma = 3$, and we need to find $P(X > 40)$, then an upper bound of 100 is easily big enough.

This would be written as: $X \sim N(3.2, 0.6^2)$

This is probably more working than you need to show in the exam – I've just included it to explain what's going on!

The next example illustrates why you must be careful using normal distribution calculations when the required answer is an integer.

Example: In a certain exam, 12% of candidates were ungraded. If the mean mark was 52 and the standard deviation was 13, and the marks were normally distributed what is the highest mark which a candidate could obtain and not gain a grade, assuming marks are integers?

Solution: Using the inverse normal function on the GDC, we find that the bottom 12% of candidates scored up to 36.7 marks. Now, 36.7 rounds to 37 – but a candidate scoring 37 would have been graded. So the highest possible mark an ungraded candidate scored is 36.

Normal Distribution: Practice Exercise

Answers
1. (a) 0.612 (b) 0.329
2. 0.48
3. 0.230
4. 78

1. For a normal distribution with mean 25 and standard deviation 3.5, find (a) $P(X < 26)$, (b) $P(24 < X < 27)$

2. The mean of a normal distribution is 1.7, and we are given that $P(X < 1.85) = 0.74$. Without using a GDC, find $P(1.55 < X < 1.85)$. *I suggest drawing a diagram.*

3. Calculate the standard deviation of the distribution in the previous question.

4. A set of exam results X is distributed normally with mean 65 and standard deviation 12. Where should the grade boundary be set such that no more than the top 15% of students gain an A grade in the exam?

μ and σ both unknown: A common situation in normal distribution questions is where you are asked to calculate the mean and standard deviation having been given two ranges and their associated probabilities. For example, you are given that the weights of an apple crop are distributed normally, and that:

- 25% of the apples weigh less than 120 g
- 15% of the apples weight more than 150 g

By using the inverse normal function, we can find the relevant Z values, and pair these with the X values (see diagram).

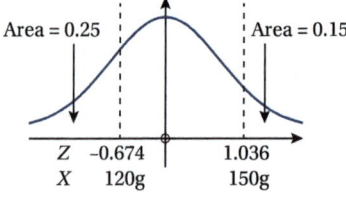

We can then substitute into the standardisation formula to form a pair of simultaneous equations:

$$-0.674 = \frac{120 - \mu}{\sigma} \text{ and } 1.036 = \frac{150 - \mu}{\sigma}$$

Solve by multiplying both sides by σ, then subtracting to eliminate μ. The solutions are: $\mu = 131.8\,\text{g}$, $\sigma = 17.5\,\text{g}$. It's a good idea to check the answer by seeing if this gives 25% of apples < 120 g.

Below I've shown all the working for a section B style question – I think it illustrates many of the strategies you need to answer longer normal distribution questions.

> Note that you will *always* be told if a distribution is normal.

A machine is set to produce bags of salt, whose weights are distributed normally, with a mean of 110 g. If the weight of a bag of salt is less than 108 g, the bag is rejected. With these settings, 4% of the bags are rejected. The settings of the machine are altered and it is found that 7% of the bags are rejected.

(a) (i) If the mean has not changed, find the new SD, correct to 3 decimal places.

What has happened is that the alteration to the machine has made it less accurate; the weights are more spread out, so more fall below 108 g. The calculator tells us that an area of 0.07 is equivalent to a standardised value of −1.4758 (use 4DP to get an accurate answer to 3DP).

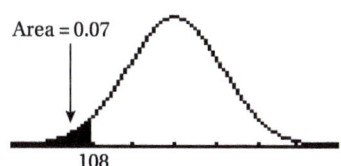

This can now be put into the standardising formula – a pivotal hinge between the calculations and the graph:

$$-1.4758 = \frac{108 - 110}{\sigma} \Rightarrow \sigma = 1.355$$

So, new standard deviation = 1.355 g

(ii) Find the value, correct to 2 decimal places, at which the mean should be set so that only 4% of the bags are rejected.

So now, accepting that we cannot improve the spread of results, we are going to increase the mean slightly (by putting more salt in each bag) and thus reduce the rejection rate. We still reject bags below 108 g.

An area of 0.04 is equivalent to $Z = -1.751$.

$$-1.751 = \frac{108 - \mu}{1.355} \Rightarrow \mu = 110.37$$

Thus, new mean = 110.37 g

The increase to the mean is small: you can see why we need to work to high accuracy.

(b) With the new settings from part (a), it is found that 80% of the bags of salt have a weight which lies between A g and B g, where A and B are symmetric about the mean. Find the values of A and B, giving your answers correct to two decimal places.

Look at the diagram. If the shaded area is 80%, then 40% is above the mean. So the total area up to B must be 90% (50 + 40).

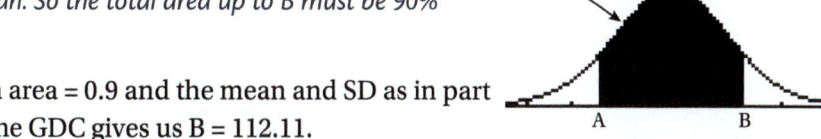

With area = 0.9 and the mean and SD as in part (a) the GDC gives us $B = 112.11$.

We can then use symmetry to find A (because it is the same distance the other side of the mean).

Thus, $A = 108.63$, $B = 112.11$

Links to other probability techniques: Probability questions do not necessarily fall into neat categories – here is a tree diagram question, there is a binomial distribution question, and that one is a Normal probability question. Quite often, a question begins by asking you to calculate a probability from, say, a Normal distribution, but then you might need to use that probability in a binomial distribution; you do need to recognise just what you are being asked.

Example: 40 players regularly train with the Griffins basketball squad. Their heights are normally distributed with mean 193 cm and standard deviation 4.8 cm.

(a) Find the probability that a member of the squad is taller than 196 cm.

(b) A team of 5 is chosen for a match. What is the probability that at least 3 of them are taller than 196 cm?

Solution: (a) Using a straightforward normal distribution calculation on the GDC, we find that P(X > 196) = 0.266.

(b) At least 3 out of 5? This is a binomial probability question with Y ~ B(5, 0.266). In other words, the normal probability we have just calculated is the "success" probability for the binomial.

P(Y ≥ 3) = 1 − P(Y ≤ 2) = 1 − 0.879 = 0.121

The length of time that John takes to drive to work can be modelled by a normal distribution with mean 16 minutes and standard deviation 2.5 minutes. He is late if his drive takes more than 20 minutes. His working week is Monday to Friday.

(a) Find the probability he is late next Monday.
(b) Find the probability that he is on time every day next week.
(c) Find the probability that he is late on at least one day for each of the next four weeks.

(a) $Z = \dfrac{20 - 16}{2.5} = 1.6$

P(Z) > 1.6 = 0.0548 (GDC)

(b) P(on time) = 1 − 0.0548 = 0.945

X ~ B(5, 0.945)

P(X = 5) = 0.945^5 = 0.754

(c) P(late on at least 1 day) = 1 − 0.754

= 0.246

P(late on at least 1 day for 4 weeks)

= 0.246^4

= 0.00364

"Next Monday" has no particular relevance, except to put the question in a real-life context. We are being asked for the probability that he is late on any one day.

Once we get to parts (b) and (c) we move to a binomial distribution. You can choose "success" either as being late or being on time. Since the question involves being on time, I've chosen to use that.

(c) Remember that P(at least one) is the same as P(1 − none). This probability then forms the basis for a new binomial distribution B(4, 0.246), where we want the probability of 4 "successes."

4.13 Continuous Distributions

Continuous random variables: Phone calls from an office last between 0 and 5 minutes. The time forms a *continuous random variable*: continuous because any value in the range can occur; random because there is no predictable pattern. The probability might be modelled by a function:

The values on the y axis are not probability, but probability density. On its own, fairly meaningless, but you've got to call the axis something!

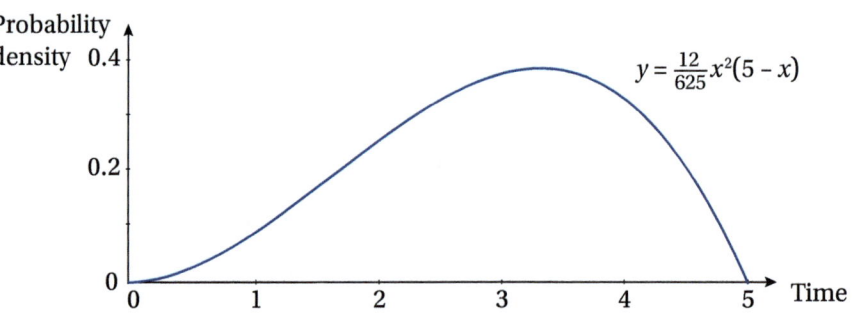

The important thing is that AREA = PROBABILITY. That accounts for the fraction $\frac{12}{625}$: it ensures that the total area under the graph is 1. In some questions this constant is written as k and you are asked to work it out. Just integrate the function over the appropriate range and put the result equal to 1.

Example: Find the probability that a call lasts between 2 and 3 minutes.

Solution: The probability is $\int_2^3 \frac{12}{625} x^2 (5-x)\, dx = 0.296$ (GDC)

Some calculators *do* have this functionality.

Mode: The mode, the most likely expected value, is the maximum point on the graph. From the calculator, this occurs when $x = 3.33$, ie 3.33 minutes is the most likely length of time for a call.

Median: The expected median will occur at the point which exactly divides the area under the graph into two equal parts. We therefore need to solve the equation $\int_0^m \frac{12}{625} x^2 (5-x)\, dx = 0.5$ for m. You cannot necessarily use your calculator for this, in which case you must do the integration "by hand." You should arrive at the equation $40m^3 - 6m^4 - 625 = 0$, which has a solution $m = 3.07$ minutes. This is the median value: you are likely to have as many phone calls less than this time as you have more.

Mean: In general, the expected mean of a continuous distribution is found using the formula $\mu = \int_{-\infty}^{\infty} x f(x)\, dx$. So, just multiply the probability density function by x and integrate. In our example, this gives the mean as 3 minutes.

The limits are always replaced by the appropriate range in any question; in this case, 0 to 5.

Variance: The formula for the expected variance is $\int_{-\infty}^{\infty} x^2 f(x)\, dx - \mu^2$. In this example this gives exactly 1, and hence the standard deviation is also 1.

Compare these formulae with those for discrete distributions on page 119.

Function definition: The probability density function will be defined in a specific format. The above example would be written as:

$$f(x) = \begin{cases} \frac{12}{625} x^2 (5-x), & 0 \le x \le 5 \\ 0, & \text{elsewhere} \end{cases}$$

A company uses Osmarc light bulbs which are changed after 3000 hours, unless they have previously failed. A probability density function for the time t hours before a bulb is changed is given as: $f(x) = \begin{cases} k e^{\frac{-t}{1200}}, & 0 < t < 3000 \\ 0 & \text{elsewhere} \end{cases}$.

(a) Find the value of the constant k.

(b) Find the probability that a bulb will last more than 2000 hours before it is changed.

(c) Find the age beyond which only 15% of Osmarc bulbs survive before being changed.

(a) $\int_0^{3000} k e^{\frac{-t}{1200}}\, dt = 1 \Rightarrow k \times 1101.5 = 1$

$k = \frac{1}{1101.5} = 9.079 \times 10^{-4}$

(b) $P(t > 2000) = \int_{2000}^{3000} 9.079 \times 10^{-4} e^{\frac{-t}{1200}}\, dt$

$= 0.116$ (GDC)

(c) $\int_T^{3000} 9.079 \times 10^{-4} e^{\frac{-t}{1200}}\, dt = 0.15$

$T = 1818$ hours (GDC)

Since the domain is $0 < t < 3000$ we can use 3000 us the upper limit in all the integrals. If we had just been told $t > 0$, the upper limit would be ∞.

(a) This is a probability distribution, so total area beneath the curve must be 1.

(b) Having found k this can be substituted into the function for the rest of the question.

(c) A more advanced use of a GDC – make sure you know how to solve an equation where the unknown is one of the integral limits. Your GDC may be able to do this directly.

Piecewise continuous distributions: We first met piecewise functions on page 110, and a probability density function can also be defined piecewise. Let's take the following as an example, and see what happens:

$$f(x) = \begin{cases} kx^2, \text{ for } 0 \leq x \leq 0.5 \\ 0.5k(1-x), \text{ for } 0.5 \leq x \leq 1 \\ 0, \text{ otherwise.} \end{cases}$$

The graph clearly has two parts which intersect when $x = 0.5$. Unfortunately, trying to find k by solving $kx^2 = 0.5k(1-x)$ when $x = 0.5$ won't work since this just yields $0.25k = 0.25k$. But we know that the total area must be 1; that is, $\int_0^{0.5} kx^2 \, dx + \int_{0.5}^1 0.5k(1-x) \, dx = 1$. Integration is simplified if you take multiplying constants out, so my working would look like this:

> I have shown plenty of working. In practice, in an exam question, the main point would be to show that the two integrals add to give 1. In paper 2 the equation could be solved directly on a GDC.

$$\int_0^{0.5} x^2 \, dx = \left[\frac{x^3}{3}\right]_0^{0.5} = 0.04167$$

$$\int_{0.5}^1 (1-x) \, dx = \left[x - \frac{x^2}{2}\right]_{0.5}^1 = 0.125$$

So $0.0416k + 0.5 \times 0.125k = 1$

$k = 9.60$

Now that we know the value of k, let's see what the graph of $f(x)$ looks like. My GDC gives me this, where I have set $0 \leq x \leq 1$ and $0 \leq y \leq 3$. If the question asked for a sketch, then this is how I would copy the display:

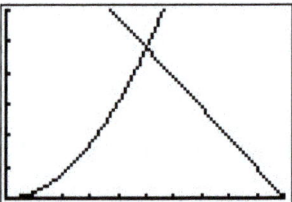

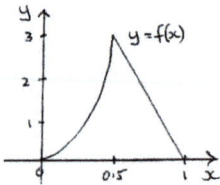

(Tip: make sure you always draw graphs and diagrams using pencil – just in case you need to amend them.)

Now try and answer the following:

(a) What is the mode of the distribution?

(b) Find P($X < 0.7$).

 Find the area from 0 to 0.5, then 0.5 to 0.7, and add them.

(c) What is the median of the distribution?

 We already know the area under the first section of the graph – is it less than or greater than 0.5? How does that help?

Answers: (a) Mode = 0.5 since $x = 0.5$ has the greatest y value.

(b) Area = $0.4 + 0.384 = 0.784$

(c) Area under first section = 0.4, so median > 0.5.

 Solve $\int_{0.5}^{m} 4.8(1-x) \, dx = 0.1$ giving $m = 0.544$

Alternatively, for part (c), we could use part of the triangle at the right hand end of the graph and make its area 0.5. The base of the triangle is $(1 - m)$, the height of the triangle is $0.5 \times 9.6 \times (1 - m)$.

4. STATISTICS AND PROBABILITY

So we need to solve:

$$\tfrac{1}{2} \times (1 - m) \times 4.8(1 - m) = 0.5$$

The solution can be found either using the equation solver on your GDC, or by rewriting the equation in quadratic form and using the relevant app.

Statistics and Probability: Long Answer Questions

Starting on the next page is a selection of Paper 2 style exam questions related to Statistics and Probability. The answers are given here, but full workings may be found on the website.

MATHEMATICS: ANALYSIS AND APPROACHES HL

1. A group of 100 students in a school are asked about whether they study History or Biology. 10 study neither, 60 study Biology, 72 study History.

    (a) n students take both subjects.

        (i) Show that $n = 42$.

        (ii) Write down the number of students who only study Biology.

    (b) One student is selected at random.

        (i) Find the probability that the student studies only one of the two subjects.

        (ii) Given that the student only studies one subject, find the probability that he studies History.

    (c) Let A be the event that a student studies History, and B be the event that a student studies Biology.

        (i) Explain why A and B are not mutually exclusive.

        (ii) Show that A and B are not independent.

    (d) There are 380 girls and 420 boys in the school. An opinion poll is to be carried out using a sample of 40 students.

        (i) Why would a quota sample not give reliable results?

        (ii) How many boys and how many girls would be surveyed if a stratified sample were to be set up?

Answers:

(a) (ii) 18

(b) (i) $\frac{48}{100}$ (ii) $\frac{30}{48}$

(c) (i) $A \cap B \neq \emptyset$ or "Some students study both subjects"

    (ii) $P(A) \times P(B) = 0.6 \times 0.72 = 0.432$. $P(A \text{ and } B) = 0.42$. $0.432 \neq 0.42$

(d) (i) Quota sampling does not produce a random sample.

    (ii) 19 girls and 21 boys.

4. STATISTICS AND PROBABILITY

2. A sample of 200 leaves is taken from a tree and their lengths, l cm are measured. The results are shown in the frequency table below.

Length	$0 < l \leq 1$	$1 < l \leq 2$	$2 < l \leq 3$	$3 < l \leq 4.5$	$4.5 < l \leq 6$	$6 < l \leq 8$	$8 < l \leq 10$
Frequency	30	38	52	35	25	13	7

(a) Calculate estimates for the mean and standard deviation of the lengths.

(b) (i) Draw a cumulative frequency diagram for the data using scales of 1 cm for 1 cm of length l on the horizontal axis, and 1 cm for 25 on the vertical axis.

(ii) Estimate the median and the interquartile range from the graph.

(iii) Use your answers to (ii) to show that a leaf measuring 8.5 cm is an outlier. Do you consider this could be an error of measurement?

(c) (i) Write down the probability that a leaf measures more than 5 cm.

(ii) Given that a leaf measures more than 5 cm, find the probability that it measures more than 8 cm.

_____

Answers:

(a) Mean = 3.09 cm, SD = 2.11 cm

(b) (ii) Median = 2.62 IQR = 4.29 − 1.53 = 2.76

(iii) Median + 1.5 × IQR = 6.76 cm ∴ 8.5 cm is an outlier. Probably not an error.

(c) (i) 0.185   (ii) 0.189

MATHEMATICS: ANALYSIS AND APPROACHES HL

3. 3 playing cards are selected at random and placed in a box; the process is then repeated with 3 more numbers placed in a second box. The two boxes are found to contain the following cards:

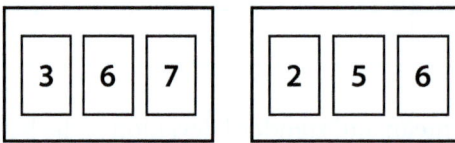

Two cards are drawn at random, one from each box.

(a) Create a list of the nine possible pairs of numbers, showing the total T in each case.

(b) Given that the selection of all pairs is equally likely, find the probability of each value of T.

(c) Find the expected value of T.

(d) George plays a game where he wins $10 if the total is less than 10, but loses $5 if the total is more than 10. What is the expected value of his winnings?

(e) How many games would George have to play before he his expected winnings were more than $45?

_____

Answers:

(a)

Pair	3,2	3,5	3,6	6,2	6,5	6,6	7,2	7,5	7,6
T	5	8	9	8	11	12	9	12	13

(b)

t	5	8	9	11	12	13
$P(T=t)$	$\frac{1}{9}$	$\frac{2}{9}$	$\frac{2}{9}$	$\frac{1}{9}$	$\frac{2}{9}$	$\frac{1}{9}$

(c) 9.67

(d) $3.33

(e) 14

4. STATISTICS AND PROBABILITY

4. A test has five questions. To pass the test, at least three of the questions must be answered correctly.

   The probability that Sammy answers a question correctly is $\frac{1}{4}$. If X is the number of questions that Sammy answers correctly,

   (a) (i) Find $E(X)$ and $Var(X)$.

   (ii) Find the probability that Sammy passes the test.

   Martha also takes the test. Let Y be the number of questions that Martha answers correctly. The following table is the probability distribution for Y.

y	0	1	2	3	4	5
$P(Y = y)$	0.34	0.49	$p - 2q$	$p - 2q$	$p - 3q$	0.01

   (b) (i) Show that $3p - 7q = 0.16$

   (ii) Given that $E(Y) = 1$ find p and q.

   (c) Find who is more likely to pass the test.

   _____

   Answers:

   (a) (i) 1.25, 0.9375

   (ii) 0.104

   (b) (ii) $p = 0.1, q = 0.02$

   (c) Martha

5. The time, in hours, that William is likely to spend on his maths assignment on a sunny Sunday is represented by the continuous random variable X which has probability density function:
   $$f(x) = \begin{cases} e - ke^{kx}, 0 \le x \le 1 \\ 0 \text{ otherwise.} \end{cases}$$

   (a) Show that $k = 1$.

   (b) Find in terms of e the mean of X.

   (c) Find the probability that Bill spends more than 30 minutes on his assignment.

   (d) (i) Find the probability that Bill spends *less* than 30 minutes on his assignment on exactly two out of three sunny Sundays?

   (ii) What assumption have you made in (d) (i)?

   _____

   Answers:

   (b) $\frac{e}{2} - 1$

   (c) 0.290

   (d) (i) 0.438

   (ii) Assume independence. The time he takes one week is not affected by the time he took the previous week.

6. A delivery driver drives the same route and measures how long it takes each time. It is found that the time, in minutes, is normally distributed with mean 122 and standard deviation 14.7

   (a) (i)   Calculate the percentage of trips which take between 110 and 130 minutes.

       (ii) Calculate how many times in 50 trips the driver will expect to take more than 130 minutes.

       (iii) Find the probability that at least 15 of the 50 trips take more than 130 minutes.

   The driver then starts a new delivery route with a mean time of 144 minutes. He finds that 9 times out of 10 his time is greater than 130 minutes.

   (b) (i)   Given that the driving time is still normally distributed, find the new standard deviation.

       (ii) What is the probability a trip will take longer than 150 minutes?

Answers:

   (a) (i)   50%

       (ii) 14.7

       (iii) 0.51

   (b) (i)   10.9

       (ii) 0.291

5. CALCULUS

Chapter 5: CALCULUS

5.1 Differentiation – The Basics

Suppose we know that the rate of inflation is 3%. This fact is useful, but would be more useful if we knew how it was changing. If its rate of change is down 0.1%/month, we can make a guess at the rate of inflation in 6 months' time. Similarly, it is useful to know we are 100 km from our destination, even more useful if we know our rate of change of distance (ie speed) is 60 kmh^{-1}. The process of finding a "rate of change function" for a given function is called differentiation. You need to know the rules for differentiating different types of function, the notation required, and the applications of differentiation.

> The *gradient* of a graph at a point represents the rate of change of the function – so differentiation gives us the gradient of a graph at any point.

Notation: When you differentiate a function, the new function (the gradient function) is called the *derived* function (or *derivative*). If the original function is $f(x)$, the derived function is written as $f'(x)$. Alternatively, if the function is written in the form $y = f(x)$, the derived function is denoted by $\frac{dy}{dx}$. Note that this doesn't mean "dy divided by dx." Effectively $\frac{d}{dx}$ is an operator whose effect is to differentiate a function of x. So we could, rather more informally, write $\frac{d(3\sin x)}{dx}$ as a shorthand for differentiating $3\sin x$.

> Don't confuse:
> $f'(x)$ Differentiated function
> $f^{-1}(x)$ Inverse function

Differentiating different types of function: You need to be able to differentiate various types of function (see table). If any functions are added or subtracted they can be differentiated independently. That is, $f(x) \pm g(x)$ differentiated is $f'(x) \pm g'(x)$.

This will not work for multiplication or division (eg to differentiate $(x + 1)(x - 2)$ you must first multiply out the brackets).

If a function is multiplied or divided by a *constant*, however, the constant just sits there: eg $2x^3$ differentiated is $2 \times 3x^2 = 6x^2$.

Also remember that functions of the form kx differentiate to give k, and that constants (which have a zero rate of change) differentiate to give 0.

Differentiating x^n: x^n differentiates to give nx^{n-1} for all $n \in \mathbb{R}$. This allows us to differentiate reciprocal and root functions. First, remember to write these functions as powers and with x in the numerator.

$f(x)$	$f'(x)$
x^n	nx^{n-1}
$\sin x$	$\cos x$
$\cos x$	$-\sin x$
$\tan x$	$\sec^2 x$
$\sec x$	$\sec x \tan x$
$\operatorname{cosec} x$	$-\operatorname{cosec} x \cot x$
$\cot x$	$-\operatorname{cosec}^2 x$
$\arcsin x$	$\frac{1}{\sqrt{1-x^2}}$
$\arccos x$	$-\frac{1}{\sqrt{1-x^2}}$
$\arctan x$	$\frac{1}{1+x^2}$
a^x	$a^x \ln a$
e^x	e^x
$\ln(x)$	$\frac{1}{x}$
$\log_a x$	$\frac{1}{x \ln a}$

MATHEMATICS: ANALYSIS AND APPROACHES HL

Examples are:

> 📱 Make sure you know how to use your GDC to find the gradient of a curve at a point.

$f(x)$	$f(x)$ rewritten	$f'(x)$	$f'(x)$ simplified
$\sqrt{x}$	$x^{\frac{1}{2}}$	$\frac{1}{2}x^{-\frac{1}{2}}$	$\frac{1}{2\sqrt{x}}$
$\frac{4}{x^2}$	$4x^{-2}$	$-8x^{-3}$	$\frac{-8}{x^3}$
$x\sqrt{x}$	$x^{\frac{3}{2}}$	$\frac{3}{2}x^{\frac{1}{2}}$	$\frac{3\sqrt{x}}{2}$
$\frac{2}{\sqrt{x}}$	$2x^{-\frac{1}{2}}$	$-\frac{1}{2} \times 2x^{-\frac{3}{2}}$	$-\frac{1}{x^{\frac{3}{2}}}$

Differentiating $\sin x$, $\cos x$ and $\tan x$: x must be in radians for these differentiations to give correct results. eg: What is the gradient of the graph of $y = x + \sin x$ when $x = 1$? $\frac{dy}{dx} = 1 + \cos x$ so when $x = 1$, the gradient is $1 + \cos 1 = 1.54$. With the calculator set in degrees, you would get 1.9998.

Differentiability at a point: If a function is *continuous* its gradient can be found for all values of x within the domain. However, discontinuous functions (such as those with vertical asymptotes, or piecewise functions) will not yield values of $\frac{dy}{dx}$ at every point. For example, at the point where two parts of a piecewise function meet, the function is not *differentiable*.

5.2 Differentiation from First Principles

What is it? It is fortunate that there are fairly simples rules and patterns for remembering how to differentiate most functions. Differentiating from first principles is a method, based on consideration of tangents, which enables us to find the differential of any function without using any rules or patterns. It requires an understanding of the idea of a *limit*.

Limits: What value does the function $f(x) = \frac{2x - 3}{x + 1}$ tend to as x gets very large? The table in the margin suggests that $f(x)$ tends towards 2; by considering the function, we can see that as x gets very large, the 3 and the 1 become insignificant, and we are indeed left with $f(x) \approx \frac{2x}{x} = 2$. This can be written as $\lim_{x \to \infty} f(x) = 2$, read as: "The limit of $f(x)$ as x tends to infinity is 2." (We also meet the idea of a limit when we consider the sum of a GP). x will never **reach** infinity, but if we think of infinity as a real place - off the map, perhaps - then the function would get to 2.

x	$f(x)$
10	1.5454...
100	1.9505...
1000	1.9950...
10000	1.9995...

A good analogy is speed. The average speed of a car which takes 2 hours to travel 140km is 70kmh⁻¹. But at a given instant, its speed might be 85 kmh⁻¹; this only means that if the car were to continue at that speed, it would travel 85km in the next hour.

Average gradient v. instantaneous gradient: The average gradient between two points is the change in y divided by the change in x. The instantaneous gradient is the gradient of the tangent at a single point.

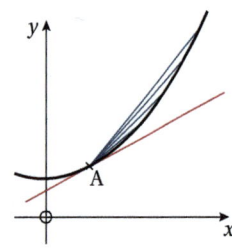

How can we calculate the gradient at a point? The diagram shows part of the graph of $y = f(x)$. We can find the gradient of the tangent at A (red line) by drawing a succession of lines (blue) to points above A, but getting closer to A, calculating the average gradients and seeing what they tend to. The process can be formalised into a general formula (note that h indicates a small distance in the x direction):

The gradient of the line from A to B is $\frac{f(x+h) - f(x)}{h}$. As B slides down the curve towards A, this value gets closer to the gradient of the tangent at A and this, of course, is the *derivative* at A. Thus we can write:

$$f'(x) = \lim_{h \to 0} \frac{f(x+h) - f(x)}{h}.$$

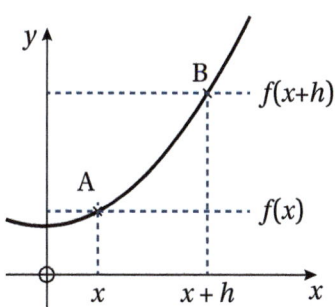

Let's see how this works in practice:

Example: Differentiate $x^2 - 2x$ from first principles.

Solution: Let's start with the top line of the formula.

$$\begin{aligned} f(x+h) - f(x) &= [(x+h)^2 - 2(x+h)] - (x^2 - 2x) \\ &= (x^2 + 2xh + h^2 - 2x - 2h) - x^2 + 2x \\ &= 2xh + h^2 - 2h \\ &= h(2x + h - 2) \end{aligned}$$

You will only be expected to differentiate polynomials from first principles.

Simplifying the top line and taking h out as a factor is important, because…

$$\begin{aligned} f'(x) &= \lim_{h \to 0} \frac{f(x+h) - f(x)}{h} \\ &= \lim_{h \to 0} \frac{h(2x + h - 2)}{h} \\ &= \lim_{h \to 0} (2x + h - 2) \\ &= 2x - 2 \end{aligned}$$

5.3 L'Hôpital's Rule

In the previous section we saw that factorising the numerator and then cancelling h allowed us to find the limit of the quotient. Without that little trick, we would have been left with a limit of $\frac{0}{0}$ which, along with a limit of $\frac{\pm\infty}{\pm\infty}$ is known as *indeterminate*. If factorisation isn't possible, we may still be able to get a result using l'Hôpital's rule which states that, if the numerator and denominator both have limits of 0 as x tends towards a, or both have limits of $\pm\infty$, then: $\lim_{x \to a} \frac{f(x)}{g(x)} = \lim_{x \to a} \frac{f'(x)}{g'(x)}$. In other words, try differentiating both functions and see whether the limit can then be found. Sometimes applying the rule results in another indeterminate limit; in which case repeated application can be tried (see last two rows of the table below).

MATHEMATICS: ANALYSIS AND APPROACHES HL

Here are some examples of l'Hôpital's rule in action:

Limit to be evaluated	Limit of $f(x)$	Limit of $g(x)$	Rule applicable?	Apply the rule	Limit
$\lim\limits_{x \to 0} \dfrac{\sin x}{x}$	0	0	Yes	$\lim\limits_{x \to 0} \dfrac{\cos x}{1}$	1
$\lim\limits_{x \to 0} \dfrac{x}{e^{2x}}$	0	1	No		0
$\lim\limits_{x \to 0} \dfrac{x}{e^{2x} - 1}$	0	0	Yes	$\lim\limits_{x \to 0} \dfrac{1}{2 e^{2x}}$	$\dfrac{1}{2}$
$\lim\limits_{x \to 1} \dfrac{1 - x^2}{\ln x}$	0	0	Yes	$\lim\limits_{x \to 1} \dfrac{-2x}{\frac{1}{x}}$	-2
$\lim\limits_{x \to +\infty} \dfrac{x^2}{e^x}$	$+\infty$	$+\infty$	Yes	$\lim\limits_{x \to +\infty} \dfrac{2x}{e^x}$	Apply again!
$\lim\limits_{x \to +\infty} \dfrac{2x}{e^x}$	$+\infty$	$+\infty$	Yes	$\lim\limits_{x \to +\infty} \dfrac{2}{e^x}$	0

(a) Show that $\ln x^x$ can be written as $\dfrac{\ln x}{\frac{1}{x}}$.

(b) Hence use l'Hôpital's rule to evaluate $\lim\limits_{x \to 0} \ln x^x$.

(c) Explain why application of the rule is valid for this function.

(a) $\ln x^x = x \ln x = \dfrac{\ln x}{\frac{1}{x}}$

(b) $\lim\limits_{x \to 0} \ln x^x = \lim\limits_{x \to 0} \dfrac{\ln x}{\frac{1}{x}}$

$= \lim\limits_{x \to 0} \dfrac{\frac{1}{x}}{-\frac{1}{x^2}}$

$= \lim\limits_{x \to 0} (-x)$

$= 0$

(c) $\lim\limits_{x \to 0} (\ln x) = -\infty$

$\lim\limits_{x \to 0} \left(\dfrac{1}{x}\right) = \infty$

∴ Application of l'Hôpital's rule is valid.

(b)(i) Note that in the third line of working it has been necessary to simplify the expression after differentiating, otherwise the limit would still be indeterminate.

5.4 The Chain Rule

The Chain Rule is used to differentiate composite functions. Consider the function $y = (4x + 3)^2$. If we write the "inner function" (ie $4x + 3$) as a single letter u, then the function becomes $y = u^2$. The chain rule shows us how the rates of change of *three* variables (as opposed to two) are connected:

$$\dfrac{dy}{dx} = \dfrac{dy}{du} \times \dfrac{du}{dx}$$

I've written a fuller explanation of the Chain Rule in an article on the website.

5. CALCULUS

We can then use the chain rule like this:

$u = 4x + 3$ $\qquad \dfrac{du}{dx} = 4$

$y = u^2$ $\qquad \dfrac{dy}{du} = 2u$

$\dfrac{dy}{dx} = \dfrac{dy}{du} \times \dfrac{du}{dx} = 2u \times 4 = 8u = 8(4x + 3)$

An alternative, informal, method is to proceed as follows:

- Take the "inner function" (in brackets) and differentiate it: $\quad 4$
- Work out the "outer function" differentiated: $\quad (...)^2 \rightarrow 2(...)$
- Multiply the two together: $\quad 8(...)$
- Fill in the brackets: $\quad 8(4x + 3)$

Here are more examples using the informal method:

$f(x) = \cos(2x - 4)$

 Inner function is $2x - 4$

 Differentiate inner $\rightarrow 2$

 Differentiate $\cos(...) \rightarrow -\sin(...)$

 Multiply $-2\sin(...)$

Result: $f'(x) = -2\sin(2x - 4)$

$f(x) = e^{x^3} = e^{(x^3)}$

 Inner function is x^3

 Differentiate inner $\rightarrow 3x^2$

 Differentiate $e^{(...)} \rightarrow e^{(...)}$

 Multiply $3x^2 \times e^{(...)}$

Result: $f'(x) = 3x^2 e^{x^3}$

$f(x) = \ln(1 + x^2)$

 Inner function is $1 + x^2$

 Differentiate inner $\rightarrow 2x$

 Differentiate $\ln(...) = \dfrac{1}{(...)}$

 Multiply $2x \times \dfrac{1}{(...)}$

Result: $f'(x) = \dfrac{2x}{(1 + x^2)}$

$f(x) = \sqrt{1 - 5x} = (1 - 5x)^{\frac{1}{2}}$

 Inner function is $1 - 5x$

 Differentiate inner $\rightarrow -5$

 Differentiate $(...)^{\frac{1}{2}} \rightarrow \frac{1}{2}(...)^{-\frac{1}{2}}$

 Multiply $-\frac{5}{2}(...)^{-\frac{1}{2}}$

Result: $f'(x) = -\frac{5}{2}(1 - 5x)^{-\frac{1}{2}}$

Chain Rule: Practice Exercise

1. Differentiate the following functions:

   (a) $\sin 2x$

   (b) $\ln(2x + 3)$

   (c) $\left(1 + \dfrac{1}{x}\right)^3$

   (d) $3\cos^2 x$

   (e) $\arctan(e^x)$

2. Given that $f(x) = e^{-x} + x$, find $f'(2)$ and the value of x for which $f'(x) = 0$.

3. Given that $f(x) = \dfrac{1}{x^2 - a}$ and $f'(2) = -1$, find the possible values of a.

4. Find $f'(x)$ if $f(x) = e^{(1 + \sin \pi x)}$.

Answers

1. (a) $2\cos 2x$

   (b) $\dfrac{2}{2x + 3}$

   (c) $\dfrac{-3}{x^2}\left(1 + \dfrac{1}{x}\right)^2$

   (d) $-6\sin x \cos x$

   (e) $\dfrac{e^x}{1 + e^{2x}}$

2. $-e^{-2} + 1,\ x = 0$

3. $a = 2$ or 6

4. $\pi \cos \pi e^{(1 + \sin \pi x)}$

MATHEMATICS: ANALYSIS AND APPROACHES HL

> Let $f(x) = \cos x$ and $g(x) = 2x^2$. Find expressions for $(g \circ f)(x)$ and $(f \circ g)'(x)$.
>
> (a) $(g \circ f)(x) = g(\cos x) = 2(\cos x)^2 = 2\cos^2 x$
>
> (b) $(f \circ g)(x) = f(2x^2) = \cos(2x^2)$
>
> $u = 2x^2 \qquad \dfrac{du}{dx} = 4x$
>
> $y = \cos u \qquad \dfrac{dy}{du} = -\sin u$
>
> $\dfrac{dy}{dx} = \dfrac{dy}{du} \times \dfrac{du}{dx} = -4x \sin u = -4x \sin(2x^2)$

Calculus questions often involve maths from any of the other areas of the syllabus – in this case we have to find a composite function and then differentiate it.

We can identify it as a chain rule differentiation because one function is contained within a second function.

> **Remember that $\ln a$ is a constant.**

Differentiating a^x: $a^x = e^{\ln a^x} = e^{x \ln a}$. This can be differentiated using the chain rule:

$$\dfrac{d}{dx}(a^x) = \ln a \times e^{x \ln a} = a^x \ln a.$$

5.5 Product and Quotient Rules

When you have to differentiate two functions multiplied together you must use the *product rule*; and when two functions are divided, you must use the *quotient rule*. If the two functions are $u(x)$ and $v(x)$ - normally shortened to u and v - then the rules are:

- Product Rule: $\qquad \dfrac{d}{dx}(uv) = u\dfrac{dv}{dx} + v\dfrac{du}{dx}$

- Quotient Rule: $\qquad \dfrac{d}{dx}\left(\dfrac{u}{v}\right) = \dfrac{v\dfrac{du}{dx} - u\dfrac{dv}{dx}}{v^2}$

> **Another quick way to remember them:**
>
> Product Rule is $uv' + vu'$
>
> Quotient Rule is $\dfrac{vu' - uv'}{v^2}$

It may be helpful to think of the rules more informally as:

> *Product Rule:*
>
> (1st fn × 2nd fn differentiated) + (2nd fn × 1st fn differentiated)

> *Quotient Rule:*
>
> $\dfrac{\text{(bottom × top differentiated)} - \text{(top × bottom differentiated)}}{\text{(bottom line squared)}}$

Note the plus sign in the product rule and the minus sign in the quotient rule. Also remember that, because of the minus sign, the order is important in the quotient rule.

When you are asked to do these more complicated differentiations, you can either write down every step in the formulae (safe but time-consuming) or you can do some of it in your head (faster, but you can go wrong). Here is an example of full working:

Example: Differentiate $y = x^2 \sin x$

Solution: $u = x^2 \qquad \dfrac{du}{dx} = 2x$

$v = \sin x \qquad \dfrac{dv}{dx} = \cos x$

$\dfrac{dy}{dx} = u\dfrac{dv}{dx} + v\dfrac{du}{dx} = x^2 \cos x + 2x \sin x$

It is possible that either (or both, if you are unlucky) of u and v are composite functions, in which case you will have to use the chain rule as well.

Example: Differentiate $f(x) = \dfrac{\sin(2x+3)}{x^2}$

Solution: $u = \sin(2x+3) \qquad \dfrac{du}{dx} = 2\cos(2x+3)$

$v = x^2 \qquad\qquad\quad \dfrac{dv}{dx} = 2x$

$$\dfrac{dy}{dx} = \dfrac{v\dfrac{du}{dx} - u\dfrac{dv}{dx}}{v^2} = \dfrac{2x^2\cos(2x+3) - 2x\sin(2x+3)}{x^4}$$

$$= \dfrac{2x(x\cos(2x+3) - \sin(2x+3))}{x^4}$$

$$= \dfrac{2x\cos(2x+3) - 2\sin(2x+3)}{x^3}$$

> Note the simplification in the last two lines. Complicated quotient rule differentiations often end up with a bit of factorisation and cancelling.

Once you have differentiated, don't forget that the end result is, as with all differentiation, however complicated, the rate of change of the original function, the gradient of the graph at any point.

There is sometimes confusion when deciding whether to use the chain rule or the product rule. If x appears twice in the function, it's probably a product rule; if not, it's either a simple differentiation or it's the chain rule.

Product and Quotient Rules: Practice Exercise

1. Differentiate, simplifying where possible:

   (a) $x^3 \ln x$

   (b) $x\sqrt{x+3}$

   (c) $\sin x \cos(2x)$

   (d) $3x^2 e^x$

2. Differentiate, simplifying where possible:

   (a) $\dfrac{x+1}{2x-1}$

   (b) $\dfrac{x^2}{e^{2x}}$

   (c) $\tan x$ *(Use one of the trig identities)*

   (d) $\dfrac{4\ln x}{x^2}$

Answers

1. (a) $x^2(1 + 3\ln x)$

   (b) $\dfrac{x}{2\sqrt{x+3}} + \sqrt{x+3}$

   (c) $-2\sin x \sin 2x + \cos x \cos 2x$

   (d) $3xe^x(x+2)$

2. (a) $\dfrac{-3}{(2x-1)^2}$

   (b) $\dfrac{2x(1-x)}{e^{2x}}$

   (c) $\sec^2 x$

   (d) $\dfrac{4(1 - 2\ln x)}{x^3}$

Notes on algebraic simplification: Questions such as those in the practice exercise above probably test your algebra more than anywhere else in the course. The issue arises: what counts as simplification, and what is just rearrangement?

The main point to make is that, on its own, factorisation isn't simplification. In question 1(a), the answer $x^2 + 3x^2 \ln x$ is equally valid. However, factorisation can lead to further simplification, particularly in algebraic fractions. Take 2(b) as an example – here's the full working:

$$f'(x) = \dfrac{e^{2x} \times 2x - x^2 \times 2e^x}{(e^{2x})^2}$$

$$= \dfrac{2xe^{2x}(1-x)}{(e^{2x})^2}$$

$$= \dfrac{2x(1-x)}{e^{2x}}$$

MATHEMATICS: ANALYSIS AND APPROACHES HL

Without the top line factorisation, the rest of the simplification could not take place. Factorisation also helps, as we shall see, when solving problems involving turning points.

The other question I am frequently asked is whether or not expressions with negative or fractional powers can be left as they are. The answer is "yes", although I prefer to get rid of them if further work (such as substituting vales for x) is going to be required. The answer I have given to 1(b) isn't the only possibility. Again, here's the working – and you'll see that we can take the working even further and write the result as a single fraction.

$$f'(x) = x \times \tfrac{1}{2}(x+3)^{-\tfrac{1}{2}} + (x+3)^{\tfrac{1}{2}} \times 1$$

$$= \frac{x}{2(x+3)^{\tfrac{1}{2}}} + (x+3)^{\tfrac{1}{2}}$$

$$= \frac{x}{2\sqrt{x+3}} + \sqrt{x+3}$$

$$= \frac{x + 2(x+3)}{2\sqrt{x+3}}$$

$$= \frac{3x+6}{2\sqrt{x+3}}$$

Any of line 1 (tidied up a bit), line 2, line 3 or line 5 would be good answers. But isn't line 5 a satisfying result compared to line 1!

Differentiating $\log_a x$: By the change of base rule, $\log_a x = \dfrac{\ln x}{\ln a}$. This looks like a quotient rule, but since $\ln a$ is a constant, it simply differentiates to give $\dfrac{1}{x \ln a}$.

Find $f'(x)$ if $f(x) = \left(\dfrac{x+3}{x-3}\right)^3$.	
$f'(x) = \dfrac{-18(x+3)^2}{(x-3)^4}$	An exciting opportunity to combine quotient and chain rules!

5.6 Second Derivative

Notation: When a function is differentiated a second time, use the notation $\dfrac{d^2y}{dx^2}$ or $f''(x)$.

Interpretation: The first derivative gives us the gradient function, so the second derivative gives us the "rate of change of gradient" function. If, for example, $f''(3) = 2$, this means that when $x = 3$, the gradient of the graph is increasing at a rate of 2 (for every increase in x of 1). It does not necessarily mean that the gradient itself is positive – only that it is increasing. This tells us about the shape of the curve. The diagram below shows what happens for various values of the first and second derivatives and covers every possible point on any curve.

Note the following:

- For a point of inflexion to occur $f''(x) = 0$, but the gradient at a point of inflexion is not necessarily 0.
- A point where $f''(x) = 0$ is not necessarily a point of inflexion. For example, $y = x^4$ has a *minimum* when $f''(x) = 0$.

- The sign of the second derivative at a turning point identifies the nature of the point: a maximum if $f''(x) < 0$, a minimum if $f''(x) > 0$. But you can also use "sign diagrams" (see the Applications of Differentiation section on page 147).

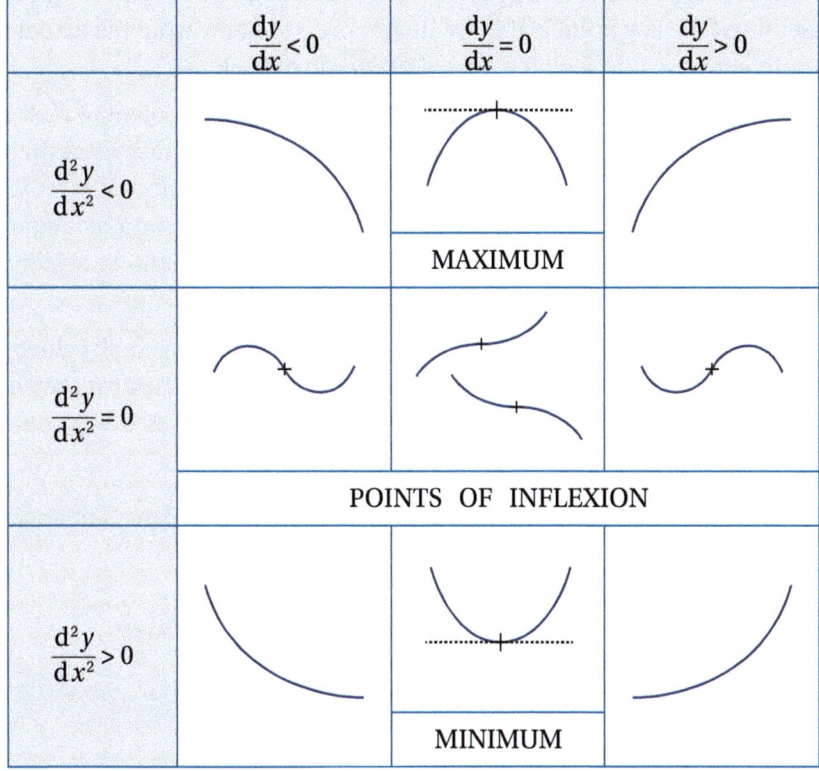

Imagine the graph is a road, and you are driving from left to right.

Right hand bends represent a decreasing gradient, so the second derivative < 0. Left hand bends represent an increasing gradient, so the second derivative > 0.

Points of inflexion occur whenever the steering wheel is momentarily straight: second derivative = 0 but this *doesn't* have to be when the gradient of the graph is 0.

Note that the parts of curves in the top row are known as "concave down" and those in the bottom row as "concave up."

Use the product rule to find and identify the stationary point on the graph of $f(x) = xe^{-x}$.

$u = x \qquad u' = 1$

$v = e^{-x} \qquad v' = -e^{-x}$

$f'(x) = uv' + vu'$

$\qquad = -xe^{-x} + e^{-x}$

For a stationary point $f'(x) = 0$

$-xe^{-x} + e^{-x} = 0$

$e^{-x}(1 - x) = 0$

$x = 1$

When $x = 1$, $y = 1 \times e^{-1} = e^{-1}$

So the stationary point is $(1, e^{-1})$ or $\left(1, \frac{1}{e}\right)$.

$f''(x) = -(-xe^{-x} + e^{-x}) - e^{-x} = xe^{-x} - 2e^{-x}$

When $x = 1$, $f''(x) = 1e^{-1} - 2e^{-1} < 0$

So the stationary point is a maximum.

We are told to use the product rule, so we can't use the GDC to find the stationary point (although if it is Paper 2 we can use the GDC to check our answer).

Note how factorisation leads straight to the answer. Also remember that exponential functions cannot ever equal zero.

To find the second derivative, note that the first term in $f'(x)$ is exactly the same as $f(x)$, with a minus sign. We've already differentiated that! Also note that we don't necessarily have to find the value of $f''(x)$ – just its sign.

5.7 Graphical Behaviour of Functions

If we are shown the graph of a function we can tell where the values of $f(x)$, $f'(x)$, and $f''(x)$ are increasing, decreasing or zero. Slightly harder, if we are given the graph of $f'(x)$ we can do the same thing. The key point is that we do not need to know what the actual function is. For example, suppose this sketch is part of the graph of $f'(x)$:

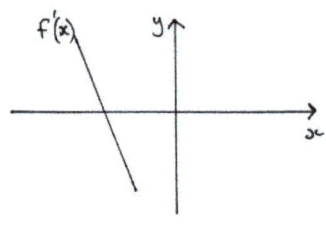

$f'(x)$ tells us the *values* of the gradient of $f(x)$. So when $f'(x) > 0$ (ie its graph is above the x-axis) the graph of the function has positive gradients – it is an *increasing* function. When $f'(x) = 0$ we've got a stationary point; and then the values of the gradient go negative, so we have a *decreasing* function.

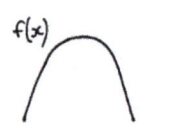

What we don't know are the actual values of the function, only its behaviour. So the next sketch shows what the graph of the function is doing, but at this stage I can't put any axes in.

A couple of examples should cover the various techniques you will need to answer exam questions.

Example: The following diagram shows a part of the graph of $y = f(x)$.

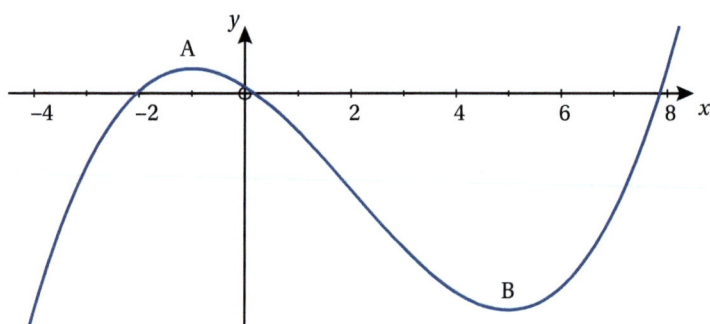

The graph has a local maximum at A, where $x = -1$, and a local minimum at B, where $x = 5$.

(a) For what values of x is f a decreasing function?

(b) Sketch the graph of $f'(x)$.

(c) Write down the following in order from least to greatest:

$f(0), f'(5), f''(-1)$.

Solution: (a) $-1 < x < 5$. (It is ***not*** decreasing at the turning points themselves).

(b) To draw the sketch, let's consider the ***values*** of $f'(x)$. Up to the point A, f is an increasing function, so the values of f' are positive – but getting less so. At A the gradient is 0, and then it is negative until B. Between A and B there is a point of inflexion; at this point the gradient reaches its largest negative value. After B the gradient is positive again, and increasing.

This leads to the following sketch:

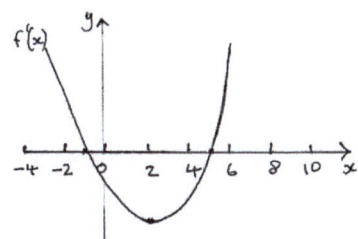

I did this sketch in pen so it would reproduce properly. But I suggest you *always* use pencil for your diagrams and graphs. So much easier to make corrections.

How do I know it's a curve? Because if it were two straight lines this would lead to a sharp point at $x = 2$, and this can't happen.

(c) $f(0)$ is positive, from original graph.

$f'(5) = 0$ because it's a stationary point

$f''(-1)$ is negative, because it's a maximum point

So, in increasing order, $f''(0), f'(5), f(0)$

Now have a look at this graph which shows the derived function, $f'(x)$, of a function which has a domain $-3 \leq x \leq 3$.

The graph of $f'(x)$ has minimum points when $x = a$ and when $x = c$, and a maximum point when $x = b$. It has zeroes when $x = a$ and $x = d$.

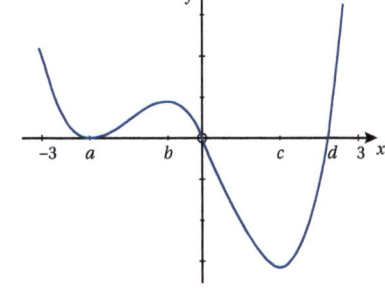

What can we deduce about the function?

- When is the function decreasing? This occurs whenever the gradient is negative, so when the graph of f' is below the axis. *Answer*: $0 < x < d$
- Where does the graph of $f(x)$ have a minimum? This will occur when the gradient is 0, so it could be at $x = a$, $x = 0$, or $x = d$. But a minimum has positive gradient to the left and negative gradient to the right; looking at the **values** of $f'(x)$ we can see that the only point which fits the criteria is where $x = 0$.
- Points of inflexion occur when $f''(x) = 0$, so at stationary points on $f'(x)$. At the same time, the gradient must have the same sign either side of the point of inflexion (ie both positive or both negative). Thus all of $x = a$, $x = b$, and $x = c$ are points of inflexion on the graph of $f(x)$.

5.8 Applications of Differentiation

Equations of tangents and normals: A tangent to a graph has the same gradient as at the point on the graph where the tangent touches, and the normal is perpendicular to the tangent. Knowing this, and the point itself, we can find the equations of the tangent and the normal. Remember that when you differentiate a function you get the *gradient function*.

MATHEMATICS: ANALYSIS AND APPROACHES HL

Example: Find the equation of the tangent to $y = 2x^2 - 4x + 3$ at the point where $x = 2$.

Solution: $\dfrac{dy}{dx} = 4x - 4$ ∴ When $x = 2$, $\dfrac{dy}{dx} = 4$

When $x = 2$, $y = 3$

$$y - y_1 = m(x - x_1)$$
$$y - 3 = 4(x - 2)$$
$$y = 4x - 5$$

> Remembering that the gradients of perpendicular lines multiply to give −1, show that the equation of the normal is $x + 4y = 14$.

Tangents and Normals: Practice Exercise

(These questions should be tried both with and without GDC).

Answers
1. $y = x - 1$
2. $(2, 11)$
3. $y = 9x + 16$
4. $y = x + 1$

1. Find the equation of the tangent to $y = x^2 \ln x$ where $x = 1$
2. T is the tangent to the curve $y = x^2 + 6x - 4$ at $(1, 3)$ and N is the normal to the curve $y = x^2 - 6x + 18$ at $(4, 10)$. Find the point of intersection of T and N.
3. Find the equation of the tangent to the curve $y = x^3 - 3x$ which is parallel to the tangent at the point $(2, 2)$.
4. Find the equation of the tangent to the curve $y = xe^{2x}$ at $x = 0$.

Maximum and minimum points: The point where a graph "turns round" can be very significant. For example, if the graph shows values of profit against selling price for a particular product, the maximum shows the selling price which leads to maximum profit.

- To find a maximum or minimum, differentiate the function then find where the gradient is 0.
- To tell which sort of point you have, use the second derivative or a sign diagram (see worked question below) if it is easier.

> Find the turning point on the graph of $y = \ln(2 + x^2)$ giving coordinates as exact values. Determine whether the turning point is a maximum or a minimum.
>
> Using the chain rule, $\dfrac{dy}{dx} = \dfrac{2x}{2 + x^2}$
>
> For a turning point, $\dfrac{dy}{dx} = 0 \Rightarrow 2x = 0 \Rightarrow x = 0$
>
> When $x = 0$, $y = \ln 2$ ∴ Turning point is $(0, \ln 2)$
>
> | x | −1 | 0 | 1 |
> |---|---|---|---|
> | $\dfrac{dy}{dx}$ | ╲ | — | ╱ |
>
> From the sign diagram $(0, \ln 2)$ is a minimum.

When finding turning points, you will often find that you are setting an algebraic fraction to equal zero. Make the numerator zero – the denominator is irrelevant.

For the sign diagram, take a value of x either side of the turning point and work out the sign of the gradient at these points. You can usually do this without working out exact values. In this case, for example, the denominator will always be positive, so just look at the sign of 2x.

You can see from the previous worked example that it is not necessary to draw a graph to find its turning points. However, make sure you can use your calculator to find maximum and minimum values: for example, find the x-coordinates of all maximums and minimums on the graph of $f(x) = \sin(1 + \sin x)$, $0 \leq x \leq 6$.

5. CALCULUS

Turning Points and Points of Inflexion: Practice Exercise

1. Find all the turning points, if any, on the following graphs, and identify them as maximums or minimums (try with and without GDC).

   (a) $y = x^3 + 6x^2 - 15x + 3$

   (b) $y = \dfrac{x}{\ln x}$

   (c) $y = \dfrac{x^2}{1+x}$

   (d) $y = 3 + \dfrac{4}{x}$

2. Given $f(x) = x^2 e^x$ where $-1 \leq x \leq 0$:

   (a) Find $f''(x)$

   (b) Solve $f''(x) = 0$

   (c) Hence find the point of inflexion on the graph of f, and use a sign diagram to confirm it is a point of inflexion.

3. Let $f(x) = e^x(2 - x^2)$

   (a) Use the product rule to show that $f'(x) = e^x(2 - 2x - x^2)$

   (b) Write down, to 3SF, the x values of the maximum and minimum points.

   (c) Find the equation of the normal to the curve where $x = 0$.

4. Prove that there are no turning points on the graph of $f(x) = 2x + \arctan x$.

Velocity and acceleration: Since velocity is rate of change of displacement, differentiating a displacement-time function will give velocity. Similarly, differentiating a velocity-time function will give acceleration (which is the rate at which velocity changes).

Let's consider the motion of a ball thrown straight up in the air and whose height h m at time t seconds is given by $h = 20t - 5t^2$, for $0 \leq t \leq 4$. We can find its velocity at any time t by differentiating: $v = \dfrac{dh}{dt} = 20 - 10t$.

The following table shows its height and velocity at different times:

t (seconds)	0	1	2	3	4
h (m)	0	15	20	15	0
v (m/s)	20	10	0	-10	-20

What does this tell us? Looking at the height, it appears to reach a maximum of 20 m before falling down and hitting the ground at 4 s. The initial velocity is 20 m/s; at 2 s its velocity is 0 m/s, confirming we are at maximum height, and then the velocity becomes negative, showing that it has reversed direction.

In the previous section we saw that to find a maximum or minimum we differentiate and set equal to 0. This is exactly what we have done here, the only difference being that the differentiated function actually has a physical meaning – velocity.

Answers

1. (a) Max(-5, 103), Min(1, -5)
   (b) Min(e, e)
   (c) Max(-2, -4), Min(0, 0)
   (d) No turning points

2. (a) $e^x(x^2 + 4x + 2)$
   (b) $x = -0.586$
   (c) $(-0.586, 0.191)$
   -ve gradient both sides

3. (b) -2.73, 0.732
   (c) $x + 2y = 4$

4. $f'(x) = \dfrac{3 + 2x^2}{1 + x^2} > 0$ since numerator and denominator are both positive.

149

An interesting point is the difference between *displacement* and *distance*. At $t = 3\,\text{s}$, the ball's displacement is 15m; this is the difference between its current position and its initial position. But the distance it has travelled is $20 + 5 = 25\,\text{m}$. We shall look at this again on page 160 in the integration section.

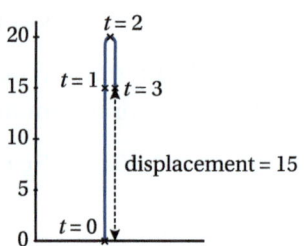

Optimisation problems: Optimisation problems boil down to this:

- find a function relating two real-life quantities;
- find the value of the first variable which leads to a maximum or minimum (the *optimum value*) of the second variable.

The second part is just the same as finding a turning point on a graph – it's often the first part which causes head scratching because it involves setting up an equation.

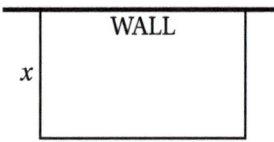

Let us suppose that a farmer has 80 m of fencing which he is going to use to enclose a rectangular sheep pen – he's going to use a stone wall as one side, and the fence to make up the other three sides of rectangle. What dimensions should the rectangle be so as to enclose the maximum area?

Let's call the width of the rectangle x. The opposite side will also be x leaving $80 - 2x$ for the length. So the area is $x(80 - 2x)$. Now it's the area we want to optimise, so:

$$A = x(80 - 2x)$$
$$= 80x - 2x^2$$
$$\frac{dA}{dx} = 80 - 4x$$

For a maximum, $80 - 4x = 0 \Rightarrow x = 20$.

So the dimensions of the rectangle are $20\,\text{m} \times 40\,\text{m}$, giving an area of $800\,\text{m}^2$. We don't need it, but I've also plotted the graph of $A = x(80 - 2x)$ to show that there is a maximum at $(20, 800)$:

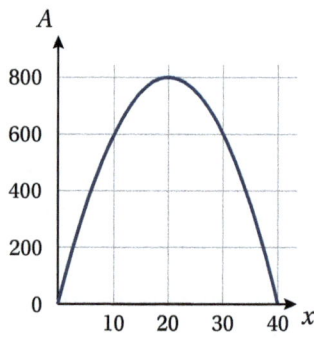

5. CALCULUS

A rectangle is bounded by the positive x-axis and the positive y-axis and the curve with equation $y = \dfrac{4-2x}{x+1}$.

(a) Show that the rectangle has maximum area when its horizontal length is $-1 + \sqrt{3}$.

(b) Find the maximum area in the form $a + b\sqrt{3}$, $a, b \in \mathbb{Z}$.

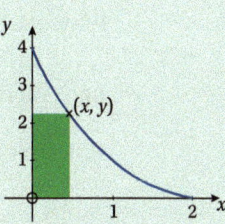

(a) Area $= xy = \dfrac{4x - 2x^2}{x+1}$

$\dfrac{dA}{dx} = \dfrac{(x+1)(4-4x) - (4x - 2x^2)}{(x+1)^2}$

$= \dfrac{(4x - 4x^2 + 4 - 4x) - (4x - 2x^2)}{(x+1)^2}$

$= \dfrac{-2(x^2 + 2x - 2)}{(x+1)^2}$

For a maximum, $x^2 + 2x - 2 = 0$

$x = \dfrac{-2 \pm \sqrt{4 - (-8)}}{2}$

$= -1 \pm \sqrt{3}$

Since $0 \leq x \leq 2$, $x = -1 + \sqrt{3}$

When $x = 0$, $\dfrac{dA}{dx} = 4$.

When $x = 2$, $\dfrac{dA}{dx} = \dfrac{-ve}{+ve}$.

So there is a maximum when $x = -1 + \sqrt{3}$

(b) Area $= \dfrac{4(-1 + \sqrt{3}) - 2(-1 + \sqrt{3})^2}{-1 + \sqrt{3} + 1}$

$= \dfrac{-4 + 4\sqrt{3} - 2 + 4\sqrt{3} - 6}{\sqrt{3}}$

$= 8 - 4\sqrt{3}$

The mathematics behind this questions is fairly straightforward, but there's a lot of detailed working to get through. The watchword here is check, and then check again, although there is help in the question itself.

(a) We have to show that the area is a maximum, so it is necessary either to find the second derivative or do a sign diagram – in this case, I opted for the latter (although I have just looked at the values of the gradient without drawing the diagram).

(b) Lots of potential traps in the working here – just be really careful with minus signs particularly.

Related rates of change: Differentiating a function tells us how fast one variable is changing compared to another. Using the chain rule, we can compare how ***three*** inter-related variables are changing. Typically, a formula connects two of them. A question might look like this:

Example: A circular oil slick is growing such that its radius is increasing at a constant rate of 5 m/hour. How fast is the area increasing when the radius is 30 m?

Solution: So, our three variables are radius, area and time. First, we write down what we are aiming for: in this case, $\dfrac{dA}{dt}$.

Now we create the chain rule formula, using the third variable: $\dfrac{dA}{dt} = \dfrac{dA}{dr} \times \dfrac{dr}{dt}$. (Imagine the dr cancelling top and bottom).

Remember that a rate of change, or just "a rate", will translate mathematically to a differential.

We know that $\frac{dr}{dt} = 5$, but what about $\frac{dA}{dr}$? For a circle, the connection between A and r is $A = \pi r^2$, so $\frac{dA}{dr} = 2\pi r$.

Let's fill it all in: $\frac{dA}{dt} = 2\pi r \times 5 = 10\pi r$, and when $r = 30$, $\frac{dA}{dt} = 300\pi$.

This gives the rate of increase of area as $942 \, \text{m}^2\text{h}^{-1}$.

A tap leaks water at $2 \, \text{cm}^3$ per second into a cone, held vertex downwards. The cone has a radius of $12 \, \text{cm}$ and a height of $18 \, \text{cm}$. Find the rate of rise of the water level when the depth is $6 \, \text{cm}$.

$\frac{dh}{dt} = \frac{dh}{dV} \times \frac{dV}{dt}$	I find a sketch usually helps:
$V = \frac{1}{3}\pi r^2 h$	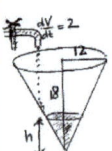
$\quad = \frac{1}{3}\pi \left(\frac{2}{3}h\right)^2 h$	
$\quad = \frac{4}{27}\pi h^3$	
So $\frac{dV}{dh} = \frac{4}{9}\pi h^2 \Rightarrow \frac{dh}{dV} = \frac{9}{4\pi h^2}$	There seem to be 4 variables (V, h, r, t) but h and r are connected since, for any height, h = 1.5r (think similar triangles). So now we can set up the chain rule using V, h and t.
Thus, $\frac{dh}{dt} = \frac{9}{4\pi h^2} \times 2 = \frac{9}{2\pi h^2}$	As before, start by writing down the differential relating to the required rate of change.
When $h = 6$, $\frac{dh}{dt} = \frac{9}{72\pi} = \frac{1}{8\pi}$	Note that we need $\frac{dh}{dV}$ so we simply find $\frac{dV}{dh}$ and take the reciprocal.

5.9 Implicit Differentiation

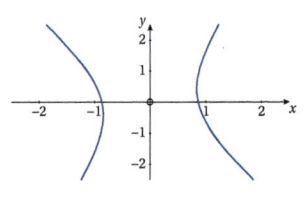

Implicit relations: A relation such as $y = 2x^2 - 3$ is called *explicit* because putting in any value of x immediately gives you a value of y. When x and y are mixed up, the relation is then called *implicit*: for example, $4x^2 + xy - y^2 = 3$ is an implicit relation. It may be possible to make x or y the subject, but not necessarily. Implicit relations are not usually functions (the graph of the relation above is illustrated), but we can still find the gradient of the curve at all points (except where the tangent is vertical). Implicit differentiation gives us a way of finding the gradient at any point on the curve.

Implicit differentiation: When you differentiate a function **with respect to** x this means that you are calculating how fast the function is changing compared to how fast x is changing. With implicit differentiation, we differentiate **each term** of the relation with respect to x. If it's already a function of x, no problem. If it's a function of y, differentiate it with respect to y then multiply by $\frac{dy}{dx}$; in other words, we find out how fast it is changing compared to y then multiply that by the rate of change of y compared to x. It sounds a bit complicated, but what it comes down to is that a term such as y^2, when differentiated with respect to x, becomes $2y\frac{dy}{dx}$.

Differentiating all the terms in the relation above, and noting that xy will have to be differentiated using the product rule, we get this:

$$8x + x\frac{dy}{dx} + y - 2y\frac{dy}{dx} = 0$$

What do we do with it now? That depends on the question:

- Find the gradient at the point $\left(\frac{\sqrt{3}}{2}, 0\right)$. Substituting the coordinates gives

$$8\frac{\sqrt{3}}{2} + \frac{\sqrt{3}}{2}\frac{dy}{dx} = 0 \Rightarrow \frac{dy}{dx} = -8$$

- Find $\frac{dy}{dx}$ in terms of y and x. To do this, we make $\frac{dy}{dx}$ the subject, just as if it were an ordinary algebraic variable. Check that this gives $\frac{dy}{dx} = \frac{-8x - y}{x - 2y} = \frac{8x + y}{2y - x}$.

It is a common mistake to leave the constant on the RHS. But 3 differentiated is 0.

Note that the gradient now depends on y as well as x.

Try this: find $\frac{dy}{dx}$ in terms of x and y given that $x\,e^{\sin y} = 4$.

Solution: $\frac{dy}{dx} = \frac{-1}{x \cos y}$.

Remember that implicit differentiation is just another weapon in the differentiation armoury. Having differentiated, you can tackle all the standard types of question such as finding the equations of tangents, finding turning points and so on.

Find the equation of the normal to the curve $2x^2 + 4y^2 = 6$ at the point A where $x = 1$ and $y < 0$.

When $x = 1$, $2 + 4y^2 = 6 \Rightarrow y = \pm 1$

Since $y < 0$, P = (1, -1)

dwrt$x$: $\quad 4x + 8y\frac{dy}{dx} = 0$

At (1, -1), $4 - 8\frac{dy}{dx} = 0 \Rightarrow \frac{dy}{dx} = \frac{1}{2}$

∴ Gradient of normal = -2

Equation of normal is $y - (-1) = -2(x - 1)$

$$y = -2x + 1$$

Although we're dealing with implicit differentiation, the basic method for finding the equation of a line is the same as ever|: find the point, find the gradient.

dwrtx is an abbreviation for "differentiate with respect to x."

In this question it would be possible to define y explicitly as a function of x. However, the implicit differentiation is actually easier since we would otherwise have to deal with positive and negative roots.

Differentiation: Practice Exercise

1. Differentiate $\operatorname{cosec} x + \tan 2x$.
2. Find the exact value of the x coordinate of the point of inflexion on the graph of $f(x) = e^{-x^2}$, $x > 0$, and justify that it is a point of inflexion.
3. Function f is defined as $f(x) = \frac{2x}{x^2 + 4}$, $x \geq a$ where $a \in \mathbb{R}$.
   (a) Find $f'(x)$, simplifying your answer.
   (b) Hence find the smallest value of a such that the inverse function f^{-1} exists.
4. Differentiate $\cos^3(3x - 2)$.
5. Differentiate $y = x^3 - 3x$ from first principles.

ANSWERS

1. $-\operatorname{cosec} x \cot x + 2\sec^2 2x$
2. $x = \frac{1}{\sqrt{2}}$
3. $\frac{8 - 2x^2}{(x^2 + 4)^2}$, 2
4. $-9\cos^2(3x - 2)\sin(3x - 2)$
6. $\frac{dy}{dx} = \frac{2^x \ln 2 - 3x^2 - 2y}{2x}$
9. -13, -15, -4; 4s
10. 0.25 cms^{-1}
11. $\frac{2\ln(x - 2)}{x - 2}$, $\frac{2 - 2\ln(x - 2)}{(x - 2)^2}$; (3, 0), (e + 2, 1)
12. ln3

MATHEMATICS: ANALYSIS AND APPROACHES HL

6. Find $\frac{dy}{dx}$ given that $2^x = x^3 + 2xy - 3$.

7. Show that the graph of $f(x) = x2^x$ has a stationary point where $x = -\frac{1}{\ln 2}$.

8. Let $y = x \arcsin x$. Show that $\frac{d^2y}{dx^2} = \frac{2-x^2}{(1-x^2)^{\frac{3}{2}}}$.

9. A particle moves in a straight line. Its displacement from the origin at time t is given by $s = t^3 - 5t^2 - 8t - 1$, $t \geq 0$. Find its position, velocity and acceleration when $t = 1$, and the time when the particle is stationary.

10. Oil is dripping onto a surface at a rate of $\frac{1}{10}\pi \, \text{cm}^3\text{s}^{-1}$. This forms a circular film with a depth of 0.1 cm. Find the rate at which the radius of the film is increasing at the moment when this radius is 2 cm. What is the rate of increase of volume? And what is the volume of the film in terms of its radius r?

11. The function f is defined as $f(x) = (\ln(x-2))^2$. Find $f'(x)$, $f''(x)$, and the coordinates of the minimum and point of inflexion on the graph of f, giving your answers in exact form.

12. Use l'Hôpital's rule to evaluate $\lim_{x \to 0} \frac{3^x - 1}{x}$.

5.10 Indefinite Integrals

Integration is sometimes called "anti-differentiation": that is, it is the reverse operation to differentiation. However, the notation is very different, and you must understand two forms – the indefinite and the definite integral.

Notation: If we just consider functions of the form ax^n, then to reverse the differentiation process we must add 1 to n then divide by the new power. For example, $4x^2$ integrated is $\frac{4}{3}x^3$. The full notation for this is: $\int 4x^2 \, dx = \frac{4}{3}x^3$. The $\int$ sign means "integrate", then comes the function you want to integrate, then dx. However, the answer is not entirely correct. If you differentiate $\frac{4}{3}x^3$ you will certainly get $4x^2$, but this will also be true if you differentiate $\frac{4}{3}x^3 + 2$, $\frac{4}{3}x^3 - 1$, and so on. In other words, when we integrate, there could be a constant at the end. Since we don't know what it is, we add a 'c' which is called "the constant of integration." So we end up with $\int 4x^2 \, dx = \frac{4}{3}x^3 + c$, and you must remember to add c to every indefinite integral – hence the word "indefinite."

> You will actually make use of the dx in integration by substitution and differential equations. Otherwise, think of it as a decorative – but necessary – piece of notation.

Integrating x^n: Generally, $\int ax^n \, dx = \frac{ax^{n+1}}{n+1} + c$ and, as with differentiation, $n \in \mathbb{R}$. There is one exception, and that is when integrating $\frac{1}{x}$. Since this is x^{-1}, the rule above would give $\frac{x^0}{0}$ and this is undefined. But when we differentiate $\ln x$ we get $\frac{1}{x}$, it follows that $\int \frac{1}{x} \, dx = \ln|x| + c$.

> The absolute value is required so that we can deal with negative values of x.

Integrating other functions: Your formula book contains a list of standard integrals, but you must also be aware of the list of derivatives. For example, $\frac{d}{dx}(\tan x) = \sec^2 x$, and it follows that $\int \sec^2 x \, dx = \tan x + c$, but you won't find this in the list of integrals.

As with differentiation, it is true that $\int f(x) + g(x) \, dx = \int f(x) \, dx + \int g(x) \, dx$ and $\int kf(x) \, dx = k\int f(x) \, dx$.

> Note that the b in $ax + b$ isn't relevant to the process.
> For example:
> $\int \cos 3x \, dx = \frac{1}{3}\sin 3x + c$.

Integrating functions of the form $f(ax + b)$: Consider what you get when you differentiate $f(x) = \sin(2x + 3)$. The "inner function" (see Chain Rule on page 140) is $2x + 3$, and this differentiates to give 2, so overall we get $f'(x) = 2\cos(2x+3)$. Now let's

154

reverse the process to find $\int \cos(2x+3)\,dx$. Since $2\cos(2x+3)$ must integrate to give $\sin(2x+3)$, it follows that $\int \cos(2x+3)\,dx = \tfrac{1}{2}\sin(2x+3) + c$. This leads to the general result: $\int f(ax+b)\,dx = \tfrac{1}{a}F(ax+b) + c$, where $F(x) = \int f(x)\,dx$. Here are some more examples:

$\int e^{3x-1}\,dx = \tfrac{1}{3}e^{3x-1}$

$\int \dfrac{1}{2x-4}\,dx = \tfrac{1}{2}\ln(2x-4) + c$

$\int (3-4x)^2\,dx = -\tfrac{1}{4} \times \tfrac{1}{3}(3-4x)^3 = -\tfrac{1}{12}(3-4x)^3 + c$

If there is a constant multiplying the function, just leave it sitting around while you do the integration – it plays no part in the proceedings, but might help simplify the final result. For example:

$\int 4\sin(2x+1)\,dx = 4 \times \left(-\tfrac{1}{2}\right)\cos(2x+1) = -2\cos(2x+1) + c$

> Given that $f(x) = (1-2x)^3$, find:
>
> (a) $f'(x)$
>
> (b) $\int f(x)\,dx$

(a) $f'(x) = -2 \times 3(1-2x)^2$ $\quad\quad\quad = -6(1-2x)^2$ (b) $\int (1-2x)^3\,dx = -\tfrac{1}{2} \times \dfrac{(1-2x)^4}{4} + c$ $\quad\quad\quad\quad\quad\quad\quad = -\dfrac{(1-2x)^4}{8} + c$	(a) Chain rule. I've used the quick method. (b) Use the $ax + b$ method, where $a = -2$.

Reversing the chain rule: Integrating $f(ax+b)$ is a specific instance of integration by reverse chain rule. More generally, if you spot an integral which is made up of a composite function multiplied by the inner function differentiated, you can guess the solution and see what happens when you differentiate it. Putting it in words is pretty incomprehensible so perhaps a diagram will help:

$\int 3x^2 \cos(x^3)\,dx$

This function is this one differentiated, so let's first try differentiating $\sin(x^3)$.

> Note that integrals using this method, and the method of the previous section, can also be found by using substitution (see page <nn>). These methods are quicker – but only work under the conditions I have set out.

Using the chain rule, we find that $f(x) = \sin(x^3) \Rightarrow f'(x) = 3x^2\cos(x^3)$ which is exactly what we want. So $\int 3x^2 \cos(x^3)\,dx = \sin(x^3) + c$.

Sometimes you have to adjust by multiplying by a constant. For example, let's examine $\int 4x(x^2+3)^3\,dx$. The overall function is a cubic, so will integrate to power 4. The inner function is x^2+3 and that differentiates to give $2x$; well, we've got $4x$ in front of the bracket, so a multiplier will sort that out. One way to tackle this is to try differentiating $(x^2+3)^4$, and this yields $8x(x^2+3)^3$. Comparing with the required integral, we conclude that $\int 4x(x^2+3)^3\,dx = \tfrac{1}{2}(x^2+3)^4 + c$.

> Look out for integrals of functions where the numerator is the derivative of the denominator. The answer will be a log.
>
> $\int \dfrac{f'(x)}{f(x)}\,dx = \ln\{f(x)\} + c$
>
> See example (e) below.

MATHEMATICS: ANALYSIS AND APPROACHES HL

Answers
(a) $\frac{1}{4}\ln(4x-3)+c$
(b) $\frac{1}{2}e^{2x}+c$
(c) $-2\cos(x^2)+c$
(d) $2(x^3-4)^{\frac{3}{2}}+c$
(e) $2\ln(x^3-2)+c$
(f) $\sqrt{2x-1}+c$
(g) $\frac{1}{4}\sin^4 x+c$
(h) $\arctan 2x+c$

Here are some $f(ax+b)$ and reverse chain rule examples for you to try:

(a) $\int \frac{1}{4x-3}\,dx$ (b) $\int e^{2x}\,dx$ (c) $\int 4x\sin(x^2)\,dx$ (d) $\int 9x^2\sqrt{x^3-4}\,dx$

(e) $\int \frac{6x^2}{x^3-2}\,dx$ (Try ln (x^3-2))   (f) $\int \frac{1}{\sqrt{2x-1}}\,dx$ (g) $\int \cos x(\sin^3 x)\,dx$;

(h) $\int \frac{1}{1+4x^2}\,dx$ Look at the list of standard integrals.

Solving gradient function equations: In some questions we are given the gradient (ie derived) function and asked to find the original function which gave rise to it. This means we must integrate the gradient function.

For example, if $f'(x) = 3x^2 - x^3$, find $f(x)$.

So, $f(x) = \int 3x^2 - x^3\,dx = x^3 - \frac{1}{4}x^4 + c$, but we will need more information to find the value of c.

Suppose we know that $f(2) = 6$ (in other words, when $x = 2$, $y = 6$). We then substitute this into the equation to get: $6 = 8 - 4 + c$, and $c = 2$. So the function we are looking for is $f(x) = x^3 - \frac{1}{4}x^4 + 2$.

As ever, exam questions may pose the question in different ways, such as:

> Full working can be found on the website

The graph of a function h passes through the point $\left(\frac{\pi}{6}, \sqrt{3}\right)$. Given that $h'(x) = 8\cos 2x$, find $h(x)$. The solution is $h(x) = 4\sin 2x - \sqrt{3}$.

5.11 Definite Integrals

> An indefinite integral:
> $\int (x^2+2)\,dx$
> A definite integral:
> $\int_1^3 (x^2+2)\,dx$

Differentiating $f(x)$ gives us a new function from which we can determine the gradient at any point on the graph of $f(x)$. Similarly, integrating $f(x)$ results in a new function from which we can find the area under the graph. To do this we must define the vertical lines which bound the area. The x values are called the *limits* and, when included, result in a *definite* integral. The integration process is the same as before, but then follow the steps necessary to evaluate the area.

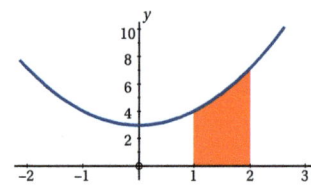

The diagram shows a shaded area bounded by the x-axis, the graph $y = x^2 + 3$, and the lines $x = 1$ and $x = 2$. The area is calculated by evaluating the definite integral $\int_1^2 x^2 + 3\,dx$.

The procedure is:

- Integrate the function, omitting the constant of integration.
- Put the result into square brackets with the limits outside.
- Substitute the limits into the integrated function (upper limit first) and subtract the two numbers. The result is the area.

> The c can be left out because it will always cancel out in the subtraction which follows.

Two important points:

- Definite integrals which evaluate an area *below* the x-axis will have negative values – although the area, of course, is still positive.
- Limits written right-to-left will also result in a negative value.

So, for our graph, the shaded area is calculated as follows:

$$A = \int_1^2 x^2 + 3 \, dx$$
$$= \left[\tfrac{1}{3}x^3 + 3x\right]_1^2$$
$$= \left(\tfrac{1}{3} \times 2^3 + 6\right) - \left(\tfrac{1}{3} \times 1^3 + 3\right)$$
$$= \tfrac{8}{3} + 6 - \tfrac{1}{3} - 3$$
$$= \tfrac{16}{3}$$

> Your GDC can evaluate definite integrals – make sure you know how.

Note how I have been careful not to take any shortcuts with the substitution and the calculation – this is particularly necessary when there are lots of minus signs flying around, as demonstrated in the next example. My motto is: you can never use too many brackets!

Example: Find the area enclosed by the x-axis, the lines $x = \tfrac{\pi}{2}$, $x = \pi$, and the graph of $y = 3\sin x - 2$.

Solution:
$$A = \int_{\tfrac{\pi}{2}}^{\pi} 3\sin x - 2 \, dx$$
$$= [-3\cos x - 2x]_{\tfrac{\pi}{2}}^{\pi}$$
$$= (-3\cos\pi - 2 \times \pi) - \left(-3\cos\tfrac{\pi}{2} - 2 \times \tfrac{\pi}{2}\right)$$
$$= (-3 \times (-1) - 2\pi) - (-3 \times 0 - \pi)$$
$$= 3 - 2\pi + \pi$$
$$= 3 - \pi$$

Now try these without a calculator:

$$\int_0^3 e^{-x} \, dx, \quad \int_3^4 (2x-6)^3 \, dx, \quad \int_{-2}^{-1} \left(\tfrac{1}{x^2} - x\right) dx, \quad \int_0^{\tfrac{\pi}{3}} 3\sin x \, dx$$

Answers:
$1 - \tfrac{1}{e^3}$, 2, 2, 1.5

5.12 Applications of Integration

Area between two curves: No problem here: simply integrate to find the area under the first curve, do the same for the second curve, then subtract to find the area between them. Except that we can make life easier for ourselves by subtracting the two functions and simplifying *before* integrating, and we then only have one integral to do. It is also probable that we will have to find where the two curves intersect, as in the following example.

MATHEMATICS: ANALYSIS AND APPROACHES HL

> Find the points of intersection of the curves $y = x^2 + 6$ and $y = 12 + 4x - x^2$ and hence find the area enclosed between them.

The curves intersect when $x^2 + 6 = 12 + 4x - x^2$

$\therefore 2x^2 - 4x - 6 = 0 \Rightarrow x^2 - 2x - 3 = 0$

$(x - 3)(x + 1) = 0 \Rightarrow x = 3$ and $x = -1$

Points of intersection are $(3, 15)$ and $(-1, 7)$.

$$\text{Area enclosed} = \int_{-1}^{3} (12 + 4x - x^2) - (x^2 + 6) \, dx$$
$$= \int_{-1}^{3} -2x^2 + 4x + 6 \, dx$$
$$= \left[-\frac{2x^3}{3} + 2x^2 + 6x\right]_{-1}^{3}$$
$$= (-18 + 18 + 18) - \left(\frac{2}{3} + 2 - 6\right)$$
$$= 21\tfrac{1}{3}$$

Here's a sketch of the two curves – but you don't need this to answer the question.

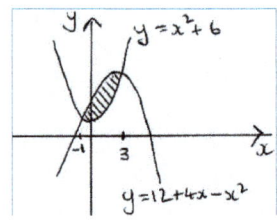

If we had subtracted the equations the other way around we would have ended up with a negative answer. But the area would still be positive.

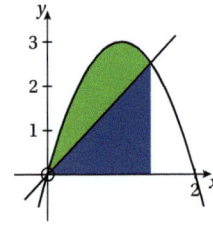

If one of the curves is actually a straight line, it may be easier to use a standard area formula. On the graph shown right, the area between the two curves is shaded green – we could find it by calculating the area under the curve, then subtracting the area of the blue triangle.

Area between a curve and the y-axis: Suppose we are asked to find the area between the curve $y = \sqrt{x - 2}$, lines $y = 1$ and $y = 3$, and the y-axis. The diagram shows the situation:

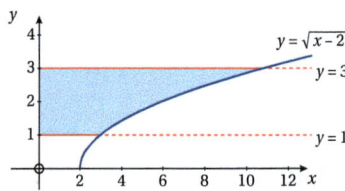

All we have to do is make sure the integral is in terms of y rather than x. First task is to rewrite the function: $x = y^2 + 2$. Now the area can be calculated as $\int_{1}^{3} y^2 + 2 \, dy = 12.66...$

Here are a couple of area questions for you to try – no GDC allowed.

1. The diagram shows part of the curve of $y = 12x^2(1 - x)$. Write down an integral which represents the area cut off by the curve and the x-axis, and find this area.

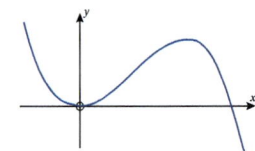

2. The diagram shows the graph of the function $y = 1 + \frac{1}{x}$. Find the value of the area of the shaded region from $y = \frac{4}{3}$ to $y = 2$ in the form $\ln k$, $k \in \mathbb{Z}^+$.

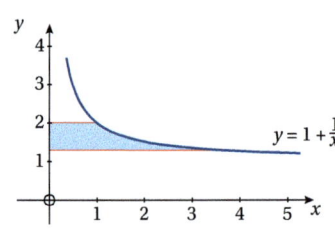

Full working on the website

Solutions: 1. $\int_{0}^{1} 12x^2(1 - x) \, dx$, 1; 2. $\ln 3$.

Volumes of revolution: The method of finding the area under a curve can be extended to calculate the volume generated when part of a curve is rotated through 360° around either the x-axis or the y-axis.

The diagram shows the shape generated when the part of the curve of $y = f(x)$ lying between $x = a$ and $x = b$ is rotated around the x-axis. Imagine a cross-section of the shape at a distance x from the origin; it will be a disc.

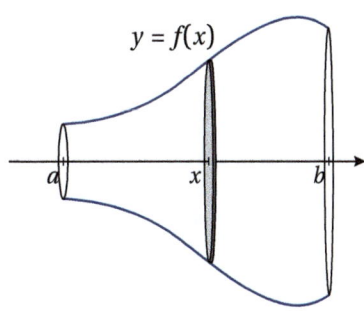

What is its volume?

- Its radius is $f(x)$
- So its cross-sectional area is $\pi\{f(x)\}^2$
- If its width is dx, its volume is $\pi\{f(x)\}^2 dx$

dx is used for a very small distance in the x direction

The overall volume will be the sum of an infinite number of such discs, and hence is found by integration.

$$V = \int_a^b \pi\{f(x)\}^2 dx$$

For clarity, this is usually written as $V = \int \pi y^2 dx$

Example: The area between the graph of $y = e^x$ and the x-axis from $x = 0$ to $x = k$ is rotated about the x-axis. Find, in terms of k, e and π, the volume generated.

Solution: Using the formula above, we get:

$$V = \int_0^k \pi(e^x)^2 dx$$

$$= \pi \int_0^k e^{2x} dx$$

$$= \pi \left[\tfrac{1}{2} e^{2x}\right]_0^k$$

$$= \pi \left(\tfrac{1}{2} e^{2k} - \tfrac{1}{2} e^0\right)$$

$$= \frac{\pi(e^{2k} - 1)}{2}$$

In the same way that you can be asked to find the area between two lines, so too you may have to find the volume generated between two curves when they are rotated 360° about either (but probably the x) axis. As a start, you will need to know where the curves intersect, as in the question that follows.

MATHEMATICS: ANALYSIS AND APPROACHES HL

> The area between the curve $y = \sqrt{2x - x^2}$ and the line $y = x$ is rotated about the x-axis through 360°. Find the volume of the solid generated.

The curves intersect when $x = \sqrt{2x - x^2}$	As with the area between two curves, the squares of the functions can be subtracted and simplified before integrating.
$x^2 = 2x - x^2$	
$2x^2 - 2x = 0$	The integral ended up with a negative value – I must have subtracted the upper curve from the lower.
$2x(x - 1) = 0$	
Solutions are $x = 0$ and $x = 1$.	
$V = \pi \int_0^1 2x^2 - 2x \, dx$	Note that in some cases where one of the curves is in fact a straight line, the question may ask you to find the volume generated by the line, which will be a cone, and only use integration for the volume generated by the curve.
$\quad = \pi \left[\frac{2x^3}{3} - x^2 \right]_0^1$	
$\quad = \pi \left(\frac{2}{3} - 1 \right)$	
$\quad = -\frac{1}{3}\pi$	
So volume $= \frac{1}{3}\pi$	

If you require the volume generated by a rotation around the y-axis, then ensure the function(s), and the limits, are in terms of y.

Kinematics problems: We first looked at kinematics problems in the differentiation section on page 149. Now we can cope with a full range of likely questions by using integration as well.

When position, velocity and acceleration are defined as functions of time, we can move from one to the other using the fact that velocity is rate of change of position, and acceleration is rate of change of velocity.

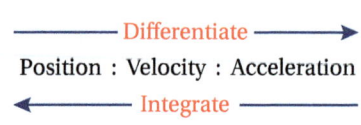

Look back to the problem involving the ball thrown up into the air, but let's suppose we were given the velocity function $v = 20 - 10t$, and not the height function. We integrate this to find position (height) and find that $h = 20t - 5t^2 + c$. Given that $h = 0$ when $t = 0$, we deduce that $c = 0$ and that $h = 20t - 5t^2$.

If we use a *definite* integral on v we will find the change in position over the given time interval, but we do need to be careful if there is a change of direction, as in fact there is at $t = 2$. This is why:

Change in position $t = 0$ to $t = 2$ is:

$$\int_0^2 20 - 10t \, dt = [20t - 5t^2]_0^2 = 20 - 0 = 20$$

Change in position $t = 2$ to $t = 3$ is:

$$\int_2^3 20 - 10t \, dt = [20t - 5t^2]_2^3 = 15 - 20 = -5$$

The negative change in position indicates we have reversed direction. So the integral from $t = 0$ to $t = 3$ will be $20 - 5 = 15$ which is its change in position compared to the start of the motion. If we want to know the total distance travelled we need to do the two separate integrals, then add the 20 upwards to the 5 downwards to get 25 m overall.

A neat way of doing this in one go using the GDC is to make use of the "absolute value" function.

See if you can set this up on your GDC:

$$\int_0^3 |20 - 10t|\, dt$$

You should get 25, any negatives having been turned into positives. (This is in your formula book.)

Some questions give velocity as a function of displacement, not time. Using the chain rule, $\frac{dv}{dt} = \frac{dv}{ds} \times \frac{ds}{dt} = v\frac{dv}{ds}$. Thus the acceleration can be expressed as $v(s)\frac{dv}{ds}$.

Example: The acceleration in ms⁻² of a particle moving in a straight line at time t seconds, $t \geq 0$, is given by the formula $a = 6t^2 + e^t$. The particle is at rest at $t = 0$.

    (a) Find an expression for v in terms of t.

    (b) What is the value of v when $t = 2.5\,\text{s}$?

Solution: (a) $v = \int 6t^2 + e^t\, dt = 2t^3 + e^t + c$

        When $t = 0$, $v = 0$ (since the particle is at rest at $t = 0$)

        $0 = 0 + 1 + c\;\;\therefore\; c = -1$

        $v = 2t^3 + e^t - 1$

    (b) When $t = 2.5$, $v = 2 \times 2.5^3 + e^{2.5} - 1 = 42.4\,\text{ms}^{-1}$

5.13 Integration by Substitution

Substituting a single letter for part of a function can make the integration of an apparently unintegrable function possible. It is important that all traces of the original letter are "removed". Suppose you are going to replace a function of x with a function of u; the following steps must be taken:

- Choose the substitution (often given to you)
- Work out $\frac{du}{dx}$ and hence the substitution for dx
- Look ahead to see if anything will cancel when you substitute
- Do the substitution, including the limits (if there are any)
- Simplify if possible, putting multiplying or dividing constants outside
- INTEGRATE (this stage is sometimes forgotten!)
- Resubstitute x if the integral was indefinite.

Watch these stages in action in the following examples:

Using the substitution $u = x^2 - 2$, integrate $x^3(x^2 - 2)^4$

$\int x^3(x^2 - 2)^4\, dx$

$= \int x^3 u^4 \frac{du}{2x}$

$= \frac{1}{2}\int x^2 u^4\, du$

$= \frac{1}{2}\int (u + 2) u^4\, du$

- Substitute $u = x^2 - 2$
- $\frac{du}{dx} = 2x \Rightarrow dx = \frac{du}{2x}$
- The x on the bottom will cancel with the x^3

Now we can multiply out and integrate to get:

$\frac{1}{2}\left(\frac{u^6}{6} + \frac{2u^5}{5}\right) + c$ which resubstitutes to $\frac{(x^2 - 2)^6}{12} + \frac{(x^2 - 2)^5}{5} + c$.

MATHEMATICS: ANALYSIS AND APPROACHES HL

The next example involves an expression on the denominator. Note how simple algebra is used to split the integral into two fractions.

Integrate $\int_0^5 \frac{x}{\sqrt{x+4}}\,dx$

$\int_0^5 \frac{x}{\sqrt{x+4}}\,dx$

$= \int_4^9 \frac{u-4}{\sqrt{u}}\,du$

$= \int_4^9 \left(\frac{u}{\sqrt{u}} - \frac{4}{\sqrt{u}}\right) du$

$= \int_4^9 (u^{\frac{1}{2}} - 4u^{-\frac{1}{2}})\,du$

$= \left[\frac{2}{3}u^{\frac{3}{2}} - 8u^{\frac{1}{2}}\right]_4^9$

- Substitute $u = x + 4$
- $\frac{du}{dx} = 1 \Rightarrow dx = du$
- When $x = 0$, $u = 4$
- When $x = 5$, $u = 9$
- We also need to use the fact that $x = u - 4$
- Because it is a definite integral we do not need to resubstitute x

Now work through the calculation without a GDC to get $4\frac{2}{3}$.

An important class of substitutions are those of type $\int \frac{1}{a^2 + x^2}\,dx$ and $\int \frac{1}{\sqrt{a^2 - x^2}}\,dx$. Here, we substitute $x = f(u)$: in the first one, $x = a\tan u$; in the second, $x = a\sin u$.

Integrate $\int \frac{1}{4 + x^2}\,dx$

$= \int \frac{1}{4 + 4\tan^2 u} 2\sec^2 u\,du$

$= \int \frac{1}{4(1 + \tan^2 u)} 2\sec^2 u\,du$

$= \int \frac{1}{2}\,du$

$= \frac{1}{2}u + c = \frac{1}{2}\tan^{-1}\frac{x}{2} + c$

- Substitute $x = 2\tan u$
- $\frac{dx}{du} = 2\sec^2 u \Rightarrow dx = 2\sec^2 u\,du$
- We need $1 + \tan^2 u = \sec^2 u$

Some for you to practise:

(Use the substitution in brackets)

$\int x(x+2)^3\,dx$ $(u = x + 2)$

$\int \frac{4x}{\sqrt{2x+1}}\,dx$ $(u = 2x + 1)$

$\int 6x^3(x^2 - 2)^6\,dx$ $(u = x^2 - 2)$

$\int \frac{2}{\sqrt{9 - x^2}}\,dx$ $(x = 3\sin u)$

Answers:

$\frac{(x+2)^5}{5} - \frac{(x+2)^4}{2} + c$

$\frac{2}{3}(2x+1)^{\frac{3}{2}} - 2(2x+1)^{\frac{1}{2}} + c$

$\frac{3(x^2-2)^8}{8} + \frac{6(x^2-2)^7}{7} + c$

$2\arcsin\left(\frac{x}{3}\right) + c$

(a) Show that the substitution $x = u^2$ transforms $\int_1^4 \frac{(1+\sqrt{x})^3}{\sqrt{x}}\,dx$ into an integral of the form $\int_a^b k(1+u)^3\,du$ where k, a and b are integers to be evaluated.

(b) Evaluate this integral.

(a) $x = u^2 \Rightarrow \frac{dx}{du} = 2u$ $\quad x = 1 \Rightarrow u = 1$

$dx = 2u\,du$ $\quad x = 4 \Rightarrow u = 2$

$\int_1^4 \frac{(1+\sqrt{x})^3}{\sqrt{x}}\,dx = \int_1^2 \frac{(1+u)^3}{u} \times 2u\,du$

$= \int_1^2 2(1+u)^3\,du$

Thus $k = 2$, $a = 1$, $b = 1$.

(b) $\int_1^2 2(1+u)^3\,du = \left[\frac{1}{2}(1+u)^4\right]_1^2$

$= \frac{1}{2} \times 3^4 - \frac{1}{2} \times 2^4$

$= 32.5$

(b) I would normally take any multipliers (2, in this case) outside the integral to make the integration easier. In this case it didn't seem worth it! The integral is of the form $F(ax + b)$, or reverse chain rule.

5.14 Integration by Parts

The typical scenario for integration by parts is when you have two functions (often of different types) multiplied together. If the two functions are called u and $\frac{dv}{dx}$, then the formula looks like this:

$$\int u \frac{dv}{dx} dx = uv - \int v \frac{du}{dx} dx.$$

> The second function is called $\frac{dv}{dx}$ because it is going to be integrated, giving v. This makes the formula simpler to understand! Note too that it is just the reverse of the product rule.

It's easier, perhaps in words:

- one of the two functions (u) **only** gets differentiated in the formula,
- the other $\left(\frac{dv}{dx}\right)$, **only** gets integrated.

It may help to think pictorially:

$$\int u \frac{dv}{dx} dx = uv - \int v \frac{du}{dx} dx$$

(Keep same, Differentiate on the right side; Integrate on the left side.)

> A simpler, informal version of the formula is:
> $\int uv' = uv - \int v'u$

Here are the steps required to find $\int 3x e^{2x} dx$

- Choose u and $\frac{dv}{dx}$ $u = 3x, \frac{dv}{dx} = e^{2x}$
- Work out $\frac{du}{dx}$ and v $\frac{du}{dx} = 3, v = \frac{1}{2}e^{2x}$
- Substitute into the formula $\int 3x e^{2x} dx = \frac{3}{2}x e^{2x} - \int \frac{3}{2} e^{2x} dx$
- Carry out the second integration $\int 3x e^{2x} dx = \frac{3}{2}x e^{2x} - \frac{3}{4} e^{2x} + c$
- Simplify if necessary $\int 3x e^{2x} dx = \frac{3}{4} e^{2x}(2x - 1) + c$

> If there are limits, leave them to the end and then put them in.

Note that the final factorisation wasn't strictly necessary, but could be very useful depending on what you've got to do next.

Bear in mind that not all products necessarily lead to integration by parts. For example, you would integrate xe^x by parts, but xe^{x^2} by reverse chain rule (or by substituting $u = e^x$).

Repeated integration by parts: Once you have worked your way through the formula, you still have an integral to do, and this could be any function, any method. Sometimes it is another integration by parts, as in the following question.

MATHEMATICS: ANALYSIS AND APPROACHES HL

> Find $\int x^2 \cos x \, dx$
>
> | | |
> |---|---|
> | $u = x^2, \frac{dv}{dx} = \cos x$. So $\frac{du}{dx} = 2x, v = \sin x$. $\int x^2 \cos x \, dx = x^2 \sin x - \int 2x \sin x \, dx$ $2\int x \sin x \, dx = 2(-x \cos x - \int -\cos x \, dx)$ $ = 2(-x \cos x + \sin x)$ $ = -2x \cos x + 2 \sin x$ So $\int x^2 \cos x \, dx = x^2 \sin x - (-2x \cos x + 2 \sin x)$ $\phantom{\text{So }\int x^2 \cos x \, dx} = x^2 \sin x + 2x \cos x - 2 \sin x + c$ | Two important points to note: Firstly, I've dealt with the second integral separately, rather than writing the whole answer out in every line (and I've taken the 2 out just to simplify the integral). Secondly, I've written everything out very carefully because of the number of minus signs flying around. This is definitely the sort of question where you need to go back and check your working. I've left the '+ c' to the end. |

Special cases: $\int x \ln x \, dx$ is clearly a candidate for integration by parts – but since we don't know how to integrate $\ln x$, we are going to have to integrate x, which seems counter-intuitive. However, it works!

Some for you to practise:
$\int x\sqrt{x-3} \, dx$
$\int x^2 \ln x \, dx$
$\int x^2 \sin x \, dx$
$\int_0^{\frac{\pi}{4}} x \cos 2x \, dx$

Answers:
$2x(x-3)^{\frac{3}{2}} - \frac{4}{5}(x-3)^{\frac{5}{2}} + c$
$\frac{x^3 \ln x}{3} - \frac{x^3}{9} + c = \frac{x^3}{9}(3 \ln x - 1) + c$
$2 \cos x + 2x \sin x - x^2 \cos x + c$
$\frac{1}{8}(\pi - 2)$

$\int x \ln x \, dx = \frac{x^2}{2} \ln x - \int \frac{x^2}{2} \times \frac{1}{x} \, dx$

$ = \frac{x^2}{2} \ln x - \int \frac{x}{2} \, dx$

$ = \frac{x^2}{2} \ln x - \frac{x^2}{4} + c$

Generally, if $\ln x$ appears in an integration by parts, take the same approach. An interesting one is $\int \ln x \, dx$ which you can rewrite as $\int 1 \times \ln x \, dx$ and then integrate by parts (to get $x \ln x - x + c$). Similarly, try integrating $\arctan x$ by the same method. Solution: $x \arctan x - \frac{1}{2} \ln(1 + x^2) + c$. Full working for both these integrals is on the website.

Another special case occurs when repeated integration by parts leads back to the integral you started with! Take the example of $\int e^x \sin x \, dx$. We'll take u as $\sin x$ and $\frac{dv}{dx}$ as e^x.

$\int e^x \sin x \, dx = e^x \sin x - \int e^x \cos x \, dx$

$ = e^x \sin x - (e^x \cos x - \int e^x (-\sin x) \, dx)$

$ = e^x \sin x - e^x \cos x - \int e^x \sin x \, dx$

At this point it looks as if we are going to go around in circles, but we can rearrange like this:

$2 \int e^x \sin x \, dx = e^x \sin x - e^x \cos x$

$\int e^x \sin x \, dx = \frac{1}{2} e^x (\sin x - \cos x) + c$

Try this method to find $\int \sin^2 x \, dx$ – start by rewriting $\sin^2 x$ as $\sin x \sin x$. Again, full working on the website.

5. CALCULUS

5.15 Integration using Partial Fractions

Splitting an unintegrable algebraic fraction into partial fractions usually makes integration possible. On page 18 we split $\frac{2x-1}{(x-2)(x-3)}$ into $\frac{-3}{x-2} + \frac{5}{x-3}$. We can therefore carry out the integration as follows:

$$\int \frac{2x-1}{(x-2)(x-3)} dx = \int \frac{-3}{x-2} + \frac{5}{x-3} dx = -3\ln(x-2) + 5\ln(x-3) + c.$$

If the question asks for an answer in the form of a single logarithm, then you can use the laws of logarithms as usual:

$$-3\ln(x-2) + 5\ln(x-3) + c$$
$$= 5\ln(x-3) - 3\ln(x-2) + c$$
$$= \ln(x-3)^5 - \ln(x-2)^3 + c$$
$$= \ln\frac{(x-3)^5}{(x-2)^3} + c$$
$$= \ln A\frac{(x-3)^5}{(x-2)^3} \text{ where } c = \ln A$$

(a) (i) Show that $\frac{2t^2}{t^2-1} \equiv 2 + \frac{a}{t^2-1}$, where a is an integer to be found.

(ii) Hence express $\frac{2t^2}{t^2-1}$ in partial fractions.

(b) (i) Using the substitution $t = \sqrt{x}$, show that $\int_4^9 \frac{\sqrt{x}}{x-1} dx = \int_2^3 \frac{2t^2}{t^2-1} dt$.

(ii) Hence show that $\int_4^9 \frac{\sqrt{x}}{x-1} dx = \ln\frac{3}{2} + 2$

(a) (i) $\frac{2t^2}{t^2-1} \equiv \frac{2(t^2-1)+2}{t^2-1} \equiv 2 + \frac{2}{t^2-1}$

So $a = 2$

(a) (ii) $\frac{2}{t^2-1} \equiv \frac{A}{t+1} + \frac{B}{t-1} \equiv \frac{-1}{(t+1)} + \frac{1}{(t-1)}$

So $\frac{2t^2}{t^2-1} \equiv 2 + \frac{1}{(t-1)} - \frac{1}{(t+1)}$

(b) (i) If $t = \sqrt{x}$ then $\frac{dt}{dx} = \frac{1}{2}x^{-\frac{1}{2}}$ When $x=4, t=2$

So $dx = 2\sqrt{x}\, dt$ When $x=9, t=3$

Substituting:

$$\int_4^9 \frac{\sqrt{x}}{x-1} dx = \int_2^3 \frac{\sqrt{x}}{t^2-1} \times 2\sqrt{x}\, dt = \int_2^3 \frac{2t^2}{t^2-1} dt$$

(b)(ii) $\int_4^9 \frac{\sqrt{x}}{x-1} dx = \int_2^3 \frac{2t^2}{t^2-1} dt$

$$= \int_2^3 2 + \frac{1}{t-1} - \frac{1}{t+1}$$
$$= [2t + \ln(t-1) - \ln(t+1)]_2^3$$
$$= (6 + \ln 2 - \ln 4) - (4 + \ln 1 - \ln 3)$$
$$= \ln\frac{3}{2} + 2$$

This is a question that involves the following techniques:
- Algebraic fractions
- Partial fractions
- Integration by substitution
- Reverse chain rule
- Rules of logarithms

None of these parts is individually hard, but it pays to be methodical and to check your working at every stage. In answering the question it's also important to let the examiner know what you are doing, especially since many of the questions ask you to "show that": all the marks in such questions must be for the working.

MATHEMATICS: ANALYSIS AND APPROACHES HL

5.16 General Methods for Integration

You may find that you can do all the various integration methods when you are told which one to use but, faced with an integration in a test or exam, it seems there are a million different ways of tackling it. This section doesn't have all the answers, but it provides some pointers for you in the form of questions you should ask yourself.

Integrals with no denominator:

- Is it a simple "reverse differentiation?"

    Examples: $\cos x$, $3x^2 - 2x + 1$, $\sec^2 x$

- Is it of the form $f(ax + b)$? The answer will be $\frac{1}{a}F(ax + b)$. Or a reverse chain rule?

    Examples: $(x - 3)^2$, $\sin(2x - 3)$, $(3x - 4)^5$, e^{4-x}, $\sqrt{(4x + 1)}$, xe^{x^2}, $4\sin^2 x \cos x$

- Might the trigonometric formulae be useful?

    Examples: $\sin^2 x$, $4\cos x \sin x$, $\tan^2 x$

- Is it two different types of function multiplied together? If so, use integration by parts – but first check it isn't a reverse chain rule.

    Examples: xe^{2x}, $x^2 \sin x$, $\ln x \, (= 1 \times \ln x)$

- If none of the above work, try integration by substitution.

    Examples: $x(2x - 3)^6$, $x^3\sqrt{x^2 - 4}$, $\sqrt{25 - x^2}$

Integrals with a denominator: The problem here is that many integrals which look just about the same will require different methods. Look at the following functions, and try to decide how you would integrate them:

(a) $\dfrac{1}{x+4}$ (b) $\dfrac{x}{x+4}$ (c) $\dfrac{1}{x^2+4}$ (d) $\dfrac{1}{x^2-4}$

(e) $\dfrac{x}{x^2+4}$ (f) $\dfrac{1}{\sqrt{4-x^2}}$ (g) $\dfrac{x}{\sqrt{4-x^2}}$

Answers:

(a) $f(ax + b)$, $\ln(x+ 4)$ \qquad (b) Substitution, $x - 4\ln(x + 4)$

(c) Formula book (or substitution), $\frac{1}{2}\arctan\frac{x}{2}$

(d) Partial fractions, $\frac{1}{4}\ln\left|\dfrac{x-2}{x+2}\right|$

(e) Reverse chain rule or substitution, $\frac{1}{2}\ln|x^2 + 4|$

(f) Formula book (or substitution), $\arcsin\frac{x}{2}$

(g) Reverse chain rule or substitution, $-(4 - x^2)^{\frac{1}{2}}$

Here are the checks you can make:

- Is the denominator to the power 1, and is the numerator the derivative of the denominator (or a simple multiple)? If so, try ln(denominator).

    Examples: $\dfrac{2}{x-3}$, $\dfrac{x}{x^2+1}$, $\dfrac{\sin x}{\cos x}$, $\dfrac{e^x}{e^x+1}$

5. CALCULUS

- Check for simple algebraic manipulation. For example, splitting up into several fractions, dividing the denominator into the numerator.

    Examples: $\dfrac{x^2 - 2x + 1}{x}$, $\dfrac{2x + 3}{x - 1}$ $\left(= 2 + \dfrac{5}{x - 1} \right)$

- Does the denominator factorise into linear factors? Try partial fractions.

    Examples: $\dfrac{2}{x^2 - 25}$, $\dfrac{3x - 1}{x^2 + 3x}$, $\dfrac{2x}{x^2 - 3x + 2}$

- Don't forget the standard integrals, particularly the inverse trigonometric functions (know what is in the formula book).

    Examples: $\dfrac{3}{x^2 + 1}$, $\dfrac{2}{\sqrt{1 - x^2}}$

- If all the above fail, rewrite the denominator in the numerator with a negative power, and see if the reverse chain rule works.

    Examples: $\dfrac{2}{(x - 1)^2}$, $\dfrac{2x}{\sqrt{x^2 + 1}}$, $\dfrac{4}{e^x}$, $\dfrac{\cos x}{\sin^3 x}$

- As a last resort, perhaps, try substitution.

    Examples: $\dfrac{x}{\sqrt{x - 1}}$, $\dfrac{x^2}{4 + x^2}$

General Integration: Practice Exercise

Here are a variety of integrals for you to try. In case you get stuck, below there are hints, and all the solutions are at the bottom. The practice integrals contain many that have been set in papers over the last few years.

Examples:

1. $\int \dfrac{4x}{1 + x^2} \, dx$
2. $\int \dfrac{x}{x + 10} \, dx$
3. $\int 4x(3x - 5) \, dx$
4. $\int \dfrac{10}{x^2 + 25} \, dx$
5. $\int \tan^2 x \, dx$
6. $\int t \sin(\tfrac{\pi}{3} t) \, dt$
7. $\int x\sqrt{2x + 3} \, dx$
8. $\int x\sqrt{x^2 - 1} \, dx$
9. $\int \cos^2 x \, dx$
10. $\int \dfrac{1 + x}{\sqrt{1 - x^2}} \, dx$
11. $\int 2t \sin(t^2) \, dt$
12. $\int \dfrac{1 + \cos x}{\sin x + x} \, dx$
13. $\int \tan 3x \, dx$
14. $\int \ln x \, dx$
15. $\int x^2 e^x \, dx$
16. $\int \dfrac{x^3}{\sqrt{1 - x^2}} \, dx$
17. $\int \dfrac{5}{x^2 - 25} \, dx$
18. $\int \dfrac{x + 7}{x^2 + 2x - 3} \, dx$

Hints:

1. Numerator is a multiple of the denominator differentiated. Try ln(denominator); adjust the constant.
2. Substitution, or divide the denominator into the numerator.
3. Just multiply out, then integrate
4. Standard integral – look it up in your formulae.
5. Which trigonometric identity contains $\tan^2 x$?
6. Product of two different types of function – by parts.
7. The outside is not the derivative of the inside; you can't multiply out; must be substitution. Try $u = 2x + 3$. Or try parts.

167

8. Outside is a multiple of the inside differentiated so you can use the reverse chain rule. You can also substitute.

9. Trig formulae. Look at the formula for $\cos 2x$.

10. Looks complicated – can we simplify algebraically? Split the numerator so you get two fractions.

11. Looks like parts – but wait: the outside is the inside differentiated, so we can use the reverse chain rule.

12. What is the connection between numerator and denominator?

13. Use the formula connecting sin, cos and tan.

14. A sneaky one. Write as $1 \times \ln x$, and use parts.

15. Parts – but you will need to do it twice.

16. Try substitution.

17. Looks like number 4, but it isn't! Split into partial fractions.

18. Also partial fractions.

Answers:

1. $2\ln(1+x^2)+c$ 2. $x - 10\ln|x+10| + c$ 3. $4x^3 - 10x^2 + c$

4. $2\tan^{-1}(\frac{x}{5}) + c$ 5. $\tan x - x + c$ 6. $-\frac{3}{\pi} t \cos(\frac{\pi}{3}t) + \frac{9}{\pi^2}\sin(\frac{\pi}{3}t) + c$

7. $\frac{1}{10}(2x+3)^{\frac{5}{2}} - \frac{1}{2}(2x+3)^{\frac{3}{2}} + c = \frac{1}{5}(x-1)(2x+3)^{\frac{3}{2}} + c$ 8. $\frac{1}{3}(x^2-1)^{\frac{3}{2}} + c$

9. $\frac{1}{4}\sin 2x + \frac{1}{2}x + c$ 10. $\sin^{-1}x - \sqrt{1-x^2} + c$ 11. $-\cos t^2 + c$

12. $\ln|x + \sin x| + c$ 13. $-\frac{1}{3}\ln|\cos 3x| + c = \frac{1}{3}\ln|\sec 3x| + c$ 14. $x\ln|x| - x + c$

15. $x^2 e^x - 2xe^x + 2e^x + c$ 16. $-(1-x^2)^{\frac{1}{2}} + \frac{1}{3}(1-x^2)^{\frac{3}{2}} + c$

17. $\frac{1}{2}\ln\left|\frac{x-5}{x+5}\right| + c$ 18. $2\ln|x-1| - \ln|x+3| + c = \ln\left|\frac{(x-1)^2}{x+3}\right| + c$

5.17 Differential Equations

Put simply, a differential equation is an equation containing a differential! In other words, you know about the rate of change of a variable, but want to end up with an equation for the variable itself. The only way to dispose of the differential is by integrating, but since the equation you end up with will contain the constant of integration, it is known as the *general solution*. If you are given specific values to substitute, you get the *particular solution*. We have already seen something like this in the section on "solving gradient function equations" on page 156.

> The particular solution is also known as the *particular integral*.

The method you use to solve a differential equation depends on its structure.

> From here on I shall use the abbreviation DE to stand for differential equation.

Variables separable: In all these examples, let's use $\frac{dy}{dx}$ as our differential. If the DE contains, or can be arranged to contain, separated functions of x and y then the method of solution is quite straightforward. We deal with three situations:

(a) $\frac{dy}{dx} = f(x)$,   (b) $\frac{dy}{dx} = g(y)$,   (c) $\frac{dy}{dx} = f(x) \times g(y)$.

In situation (a) we simply integrate both sides with respect to x. In other words, the general solution is $y = \int f(x)\,dx$.

Situation (b) is not so straightforward since we can't integrate $\int g(y)\,dx$. However, imagine that we can separate dy and dx, then (b) can be rewritten as $\int \frac{1}{g(y)}\,dy = \int 1\,dx$ and therefore the general solution is $x = \int \frac{1}{g(y)}\,dy$, which may have to be rearranged if y is to be the subject.

This looks like a fiddle but, it does in fact work!

Situation (c) works just like (b): this time we get $\int \frac{1}{g(y)}\,dy = \int f(x)\,dx$. So to get to the general solution we must deal with two integrals.

Here's an example of each of (a), (b) and (c):

$\frac{dy}{dx} = x + e^x$	$\frac{dy}{dx} = 2y + 3$	$x\frac{dy}{dx} - y^2 = 0$
$y = \int x + e^x\,dx$	$\int \frac{1}{2y+3}\,dy = \int 1\,dx$	$x\frac{dy}{dx} = y^2$
$y = \frac{x^2}{2} + e^x + c$	$\frac{1}{2}\ln|2y+3| = x + c$	$\int \frac{1}{y^2}\,dy = \int \frac{1}{x}\,dx$
	$\ln|2y+3| = 2x + c$	$-\frac{1}{y} = \ln x + c$
	$2y + 3 = e^{2x+c}$	$y = -\ln(Ax)$
	$y = \frac{Ae^{2x} - 3}{2}$	

In (b), note how the $+c$ becomes the multiplicative constant A.

And in (c), replace c by $\ln A$ in the 4th line to get $\ln(Ax)$

You will find that, after rearrangement, you could be faced with using any of the integration methods we have met. The following exercise is also good practice for integral spotting!

Variables Separable: Practice exercise

Find the general solutions of the following DEs.

1. $\frac{dy}{dx} = \sec^2 y$

   Which function is the reciprocal of sec? You will also need a double angle formula.

2. $(x^2 - 1)\frac{dy}{dx} = y$

   Look out for partial fractions.

3. $x\frac{dy}{dx} = y + xy$

   Factorise the RHS. Then everything x to the right, everything y to the left.

4. $\frac{dy}{dx} = \frac{1}{\sqrt{xy}}$

   Start by splitting up the square root so you can then separate the variables.

5. $(\sin x - \cos x)\frac{dy}{dx} = y(\cos x + \sin x)$

   Once you have a function of x on the RHS, it looks like trigonometrical formula. But always check first – is the numerator the derivative of the denominator?

6. $\sin y \frac{dy}{dx} = x^3 \sec^3 y$

   It will probably help to rewrite $\sec^3 y$ in terms of $\cos y$.

ANSWERS

1. $x = \frac{1}{2}y + \frac{1}{4}\sin 2y + c$
2. $y = A\sqrt{\frac{x-1}{x+1}}$
3. $y = Axe^x$
4. $y = (3x^{\frac{1}{2}} + c)^{\frac{2}{3}}$
5. $y = A(\sin x - \cos x)$
6. $\cos^4 y = -x^4 + c$

MATHEMATICS: ANALYSIS AND APPROACHES HL

If we are given an x value and a y value which satisfies the equation, then we can work out the value of the constant, and hence the particular solution. The following question sets a DE in a practical context.

> Liquid is being poured into a container at a constant rate of $10\,\text{cm}^3\text{s}^{-1}$ and is leaking out a rate proportional to the volume of liquid in the container.
>
> (a) Explain why the volume of liquid, $V\,\text{cm}^3$, at time t seconds satisfies the differential equation $\dfrac{dV}{dt} = 10 - kV$ where k is a positive constant.
>
> (b) Given that the container is filling at a rate of $6\,\text{cm}^3\text{s}^{-1}$ when it $V = 20\,\text{cm}^3$, find the value of k.
>
> The container is initially empty.
>
> (c) Solve the differential equation, making V the subject.
>
> (d) Find the volume of liquid in the container 15 seconds after the start.

(a) Every second $10\,\text{cm}^3$ goes in, and kV leaks out, so rate of increase is $10 - kV$, where k is the constant of proportionality.

(b) $6 = 10 - k \times 20 \Rightarrow k = 0.2$

(c) $\dfrac{dV}{dt} = 10 - 0.2V$

$$\int \dfrac{5}{50 - V}\,dV = \int 1\,dt$$

$$-5\ln(50 - V) = t + c$$

$$50 - V = e^{-\frac{t}{5}+c}$$

$$V = 50 - Ae^{-\frac{t}{5}}$$

When $t = 0$, $V = 0$

$0 = 50 - A \Rightarrow A = 50$

So solution is $V = 50 - 50e^{-\frac{t}{5}}$

(d) When $t = 15$, $V = 50 - 50e^{-3} = 47.5\,\text{cm}^3$

(c) You'll see that I've multiplied through by 5 so that all the numbers are integers. This makes working much easier.

In the fourth line of working The c actually becomes $-\frac{1}{5}c$, but it's still a constant, so I just leave it as c.

(d) We could also have been asked a question about limits here. Since the term in e is tending towards zero, the volume is tending towards 50. At this point, the amount leaking out is the same as the amount pouring in.

Homogeneous differential equations: If a DE can be reduced to the form $\dfrac{dy}{dx} = f\!\left(\dfrac{y}{x}\right)$ it is said to be *homogeneous*. Such equations can be solved by the substitution $y = vx$. For example, let's solve the DE $\dfrac{dy}{dx} = \dfrac{x + 2y}{x}$.

> This DE is homogeneous because it can be rewritten as $\dfrac{dy}{dx} = 1 - 2\!\left(\dfrac{y}{x}\right)$.

- First we use the product rule to write $\dfrac{dy}{dx} = v + x\dfrac{dv}{dx}$.

- Now do the substitution: $v + x\dfrac{dv}{dx} = \dfrac{x + 2vx}{x}$.

- Then we simplify and, hopefully, end up with a variables separable DE.

$$v + x\dfrac{dv}{dx} = 1 + 2v$$

$$\int \dfrac{1}{1 + v}\,dv = \int \dfrac{1}{x}\,dx$$

$$\ln|1 + v| = \ln|x| + c$$

$$\ln|1+v| = \ln|Ax| \text{ (where } \ln A = c\text{)}$$
$$1 + v = Ax$$

- Now resubstitute: $1 + \frac{y}{x} = Ax$
$$y = Ax^2 - x$$

Now try finding the general solution of $x\frac{dy}{dx} = 3x - 2y$, and the particular solution given that $y = \frac{3}{4}$ when $x = 1$.

Answers: $y = x - \frac{A}{x^2}, y = x - \frac{1}{4x^2}$

> Full working on the website

Using the integrating factor: If a DE is of the form $\frac{dy}{dx} + P(x)y = Q(x)$ then it can be solved by multiplying through by an *integrating factor*. In this case, the integrating factor is $e^{\int P(x)dx}$. Let's follow this through:

$$e^{\int P(x)dx}\frac{dy}{dx} + e^{\int P(x)dx}P(x)y = e^{\int P(x)dx}Q(x)$$

Not too encouraging, at first sight. But if you use implicit differentiation, you will find that the LHS is equivalent to $\frac{d}{dx}(ye^{\int P(x)dx})$. So the equation simplifies to:

$$\frac{d}{dx}(ye^{\int P(x)dx}) = e^{\int P(x)dx}Q(x)$$

and by integrating both sides, we finally get:

$$ye^{\int P(x)dx} = \int e^{\int P(x)dx}Q(x)\,dx$$

Again, this looks pretty horrific, but by working through the functions carefully, we should end up with a "doable" integral. I would set my working out like this:

> It helps to call the integrating factor I, in which case the formula can be remembered as:
> $yI = \int IQ\,dx$

Example: Solve the differential equation $\frac{dy}{dx} + \frac{3y}{x} = x$

Solution: Look back to the general equation at the start of this section. We can see that $P(x) = \frac{3}{x}$ and $Q(x) = x$. So, $e^{\int P(x)dx} = e^{\int \frac{3}{x}dx} = e^{3\ln x} = e^{\ln x^3} = x^3$.

Now substitute:
$$yx^3 = \int x^3 \times x\,dx$$
$$yx^3 = \frac{x^5}{5} + c$$
$$y = \frac{x^2}{5} + \frac{c}{x^3}$$

> Of course, you may not be so fortunate to get such a simple integral!

Now try these: (a) $\frac{dy}{dx} - 2xy = e^{x^2}$ (b) $(1 + x^2)\frac{dy}{dx} + 2xy = \frac{4}{1 + x^2}$ (Hint: First divide through both sides by $(1 + x^2)$. Why?)

Answers: $y = e^{x^2}(x + c); y = \frac{4\tan^{-1}x + c}{(1 + x^2)}$.

> Full working on the website.

Euler's method: Although you have learnt some methods for solving differential equations algebraically most cannot, in fact, be solved. Euler's method is a *numerical method* which can be used instead – but only for finding a particular solution. Since it is not algebraic it cannot be used to find the general solution. It uses an *iterative* formula, which means that the result from the formula at each stage is put back into the same formula, each new result becoming ever closer to the solution. The iterations are stopped when the desired x value is reached.

Euler's method is used to solve the DE $\frac{dy}{dx} = f(x,y)$: that is, we will find the value of y for a given value of x.

The iterative formula is in your formula book, and is $y_{n+1} = y_n + hf(x_n, y_n)$. As with many of these formulae, it looks more complicated than it really is! Here's how it works:

- h is the *step size*, the small amount by which we step through values of x until we reach the one we want.
- n is the number of the iteration, starting with $n = 0$.
- So we start with our known values, x_0 and y_0, and substitute them into f to get $f(x_0, y_0)$. Multiply this by h and add to y_0. This gives us y_1.
- Now $x_1 = x_0 + h$ and we start all over again with our new values x_1 and y_1.

The safest way to answer an Euler's method question is to construct a table. Let's use the method to find y when $x = 0.3$ given that $\frac{dy}{dx} = xy$, and when $x = 0$, $y = 1$. We shall use a step length of 0.1, ie $h = 0.1$.

The table shows how Euler's method works. However, in an exam, having written down the formula, you should be able to use your GDC to generate the successive terms – there is no need to construct the table.

Step (n)	x_n	y_n	$y_n' = f(x_n, y_n)$	hy_n'	$y_{n+1} = y_n + hy_n'$
0	0	1	0	0	1
1	0.1	1	0.1	0.01	1.01
2	0.2	1.01	0.202	0.0202	1.0302
3	0.3	1.0302	0.30906	0.030906	1.061106

The solution is therefore $y \approx 1.0611$ when $x = 0.3$. In this case, because the variables are separable, we can also compare this approximation with the exact answer:

$$\frac{dy}{dx} = xy$$

$$\int \frac{1}{y} dy = \int x \, dx$$

$$\ln|y| = \frac{x^2}{2} + c$$

$$y = Ae^{\frac{x^2}{2}}$$

The accuracy can always be improved by decreasing the step length.

When $x = 0$, $y = 1$, so $A = 1$, and therefore the particular solution is $y = e^{\frac{x^2}{2}}$. When $x = 0.3$, $y = e^{\frac{0.09}{2}} = 1.04603$. The error in Euler's method is just 1.4%.

Now try using Euler's method with $h = 0.2$ to find $f(2)$ given that $f'(x) = x - y^2$ and that $f(1) = 0$. Answer: $y = 1.10033$.

5.18 Maclaurin Series

Apart from polynomials and rationals, the values of most functions cannot be calculated arithmetically in a single step. For example, if $x = 0.1$, what are the values of $\sqrt{1+x}$, $\sin x$, e^{2x}? Maclaurin's Theorem gives us a way to find polynomial curves which approximate to these functions around the point where $x = 0$. In each case we can generate a series, and greater accuracy is achieved by taking more terms of the series: effectively, the higher the degree of the polynomial approximation, the better the fit.

Maclaurin's theorem states that:

$$f(x) = f(0) + xf'(0) + \frac{x^2}{2!}f''(0) + \frac{x^3}{3!}f'''(0) + \ldots$$

When working out a series, it is best to start by finding the derivatives. Here is the working for the Maclaurin series for e^x up to the term in x^3.

$$f(x) = e^x \qquad f(0) = 1$$
$$f'(x) = e^x \qquad f'(0) = 1$$
$$f''(x) = e^x \qquad f''(0) = 1$$
$$f'''(x) = e^x \qquad f'''(0) = 1$$

So
$$e^x \approx 1 + x \times 1 + \frac{x^2}{2!} \times 1 + \frac{x^3}{3!} \times 1$$
$$\approx 1 + x + \frac{x^2}{2} + \frac{x^3}{6}$$

Substituting $x = 0.2$ gives $e^{0.2} \approx 1.22133$

To 5DP $e^{0.2}$ is in fact 1.22141

You will find Maclaurin series for e^x, $\sin x$, $\cos x$, $\ln(1 + x)$. The series for $(1 + x)^n$ is the same as the binomial expansion (see page 15). So what might you be asked to do in an exam question?

Substitution: By replacing x with $f(x)$ a new series can be generated. For example, by replacing x with $2x$ the series for $\ln(1 + 2x)$ is $2x - 2x^2 + \frac{8x^3}{3} - \ldots$

Combining series: We can find the series of, for example, $e^x \sin x$ by multiplying the individual series together. Be careful not to overdo the working – if you only need to go up to the term in x^3, select the appropriate terms to multiply. For example, to calculate the series for $e^x \sin x$ up to the term in x^3:

$$e^x \sin x \approx \left(1 + x + \frac{x^2}{2} + \frac{x^3}{6}\right)\left(x - \frac{x^3}{6}\right)$$
$$\approx x - \frac{x^3}{6} + x^2 + \frac{x^3}{2}$$
$$= x + x^2 + \frac{x^3}{3}$$

I thought you might be interested to see how the graphs of each successive polynomial fit against the original function. Original is red, the linear is blue, quadratic is purple and cubic is green – which actually covers the red quite a bit!

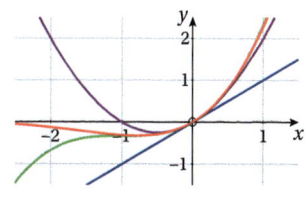

Example: By substituting $(-x)$ for x, find the Maclaurin series for $\ln(1 - x)$ up to the term in x^5. Hence find the Maclaurin series for $\ln\sqrt{\frac{1+x}{1-x}}$ up to the term in x^5.

Solution:
$$\ln(1 + x) = x - \frac{x^2}{2} + \frac{x^3}{3} - \frac{x^4}{4} + \frac{x^5}{5}$$

$$\ln(1 - x) = (-x) - \frac{(-x)^2}{2} + \frac{(-x)^3}{3} - \frac{(-x)^4}{4} + \frac{(-x)^5}{5}$$

$$= -x - \frac{x^2}{2} - \frac{x^3}{3} - \frac{x^4}{4} - \frac{x^5}{5}$$

$$\ln\sqrt{\frac{1+x}{1-x}} = \tfrac{1}{2}(\ln(1+x) - \ln(1-x))$$

$$= \tfrac{1}{2}\left(\left(x - \frac{x^2}{2} + \frac{x^3}{3} - \frac{x^4}{4} + \frac{x^5}{5}\right) - \left(-x - \frac{x^2}{2} - \frac{x^3}{3} - \frac{x^4}{4} - \frac{x^5}{5}\right)\right)$$

$$= \tfrac{1}{2}\left(2x + \frac{2x^3}{3} + \frac{2x^5}{5}\right)$$

$$= x + \frac{x^3}{3} + \frac{x^5}{5}$$

MATHEMATICS: ANALYSIS AND APPROACHES HL

Differentiating and integrating a series: Differentiation or integration of a series, term by term, can often lead to other series, or interesting proofs.

For example, if we differentiate the series for e^x:

$$e^x = 1 + x + \frac{x^2}{2!} + \frac{x^3}{3!} + \frac{x^4}{4!} + \ldots$$

$$\frac{d}{dx}(e^x) = 0 + 1 + \frac{2x}{2!} + \frac{3x^2}{3!} + \frac{4x^3}{4!} + \ldots$$

$$= 1 + x + \frac{x^2}{2!} + \frac{x^3}{3!} + \ldots$$

$$= e^x$$

…thus proving that $\frac{d}{dx}(e^x) = e^x$!

(a) By use of an appropriate Maclaurin series, find the series for $\frac{\sin x}{x^2}$ up to the term in x^3.

(b) Hence find an approximation for $\int_{0.1}^{0.2} \frac{\sin x}{x^2} dx$ giving your answer to 5DP.

(a) $\sin x = x - \frac{x^3}{3!} + \frac{x^5}{5!}$

$\therefore \frac{\sin x}{x^2} = \frac{1}{x} - \frac{x}{6} + \frac{x^3}{120}$

(b) $\int_{0.1}^{0.2} \frac{\sin x}{x^2} dx = \int_{0.1}^{0.2} \frac{1}{x} - \frac{x}{6} + \frac{x^3}{120} dx$

$= \left[\ln|x| - \frac{x^2}{12} + \frac{x^4}{480} \right]_{0.1}^{0.2}$

$= (-1.612768) - (-2.303418)$ (GDC)

$= 0.69065$

(a) Clearly the appropriate series is the one for sin x. Make sure you are aware which series are in your formula book.

Knowing we have to divide by x^2, we take the sin x series up to the x^5 term.

(b) A quick way to do the calculation here is to enter the integral function onto your GDC, then use the function value facility. You should know how to use your GDC to carry out a calculation such as $f(0.2) - f(0.1)$.

> Throughout the working, $f(x)$ and y are interchangeable, as are $f'(x)$ and $\frac{dy}{dx}$.

Developing a series from a differential equation: Suppose we are given the DE $\frac{dy}{dx} = x - 4xy$ with initial condition $f(0) = 1$. Our aim is to find a Maclaurin series for $f(x)$. We need to go back to the Maclaurin Theorem formula:

$$f(x) = f(0) + xf'(0) + \frac{x^2}{2!}f''(0) + \frac{x^3}{3!}f'''(0) + \ldots$$

We know that $f(0) = 1$ and we know that $f'(0) = 0 - 4 \times 0 \times 1 = 0$. Now we need to do some differentiating to find the higher derivatives. To find $f''(x)$ we shall differentiate $f'(x)$ and so on. $f''(x) = \frac{d}{dx}(x - 4xy) = 1 - 4y - 4x\frac{dy}{dx}$ (using the product rule and implicit differentiation). So $f''(0) = 1 - 4 - 0 = -3$. Carrying on:

$$f'''(x) = 0 - 4f'(x) - 4f'(x) - 4xf''(x) = -8f'(x) - 4xf''(x)$$

$$f^{iv}(x) = -8f''(x) - 4f''(x) - 4xf'''(x) = -12f''(x) - 4xf'''(x)$$

These give $f'''(0) = 0$ and $f^{iv}(0) = 36$

So (finally!) we find $f(x) = 1 + 0 + \frac{x^2}{2} \times (-3) + 0 + \frac{x^4}{24} \times (36) + \ldots$

$$= 1 - \frac{3x^2}{2} - \frac{3x^4}{2} + \ldots$$

5. CALCULUS

Calculus: Long Answer Questions

The answers to these section B style questions are given; full working can be found on the website at www.peakib.com.

MATHEMATICS: ANALYSIS AND APPROACHES HL

1. Let $f(x) = \dfrac{2x}{x^2 + 4}$ where $x > 0$.

    (a) Show that $f'(x) = \dfrac{8 - 2x^2}{(x^2 + 4)^2}$

    (b) The graph of f has a maximum point. Find its coordinates.

    (c) Find $\int \dfrac{2x}{x^2 + 4}\, dx$

    (d) The area enclosed by the x-axis, the graph of f, and the lines $x = 1$ and $x = a$ is $\ln \dfrac{13}{5}$. Find the value of a.

    _____

    Answers:

    (b) $(2, 0.5)$ *Note the domain of f*

    (c) $\ln(x^2 + 4) + c$

    (d) $a = 3$

2. The function f has its derivative defined by $f'(x) = 3x^2 - px - 12$, where p is a constant.

    (a) Find $f''(x)$.

    (b) Given that the graph of f has a point of inflexion when $x = 1$, show that $p = 6$.

    (c) Find $f'(-2)$.

    The point $P(-2, 3)$ lies on the graph of f.

    (d) Find the equation of the tangent to the curve of f at P, giving your answer in the form $ax + by = c$.

    (e) Find $f(x)$.

    _____

    Answers:

    (a) $f''(x) = 6x - p$

    (b) $6 \times 1 - p = 0 \Rightarrow p = 6$

    (c) $f'(-2) = 12$

    (d) $y - 12x = 27$

    (e) $f(x) = x^3 - 3x^2 - 12x - 1$

5. CALCULUS

3. A circular oil slick on the surface of the sea had a radius of 45 m when first observed, and its radius was increasing at 0.8 m/min. At a time t minutes later, its radius is r metres.

    If left untreated, its radius will increase at a rate which is inversely proportional to the square of its radius.

    (a) (i) Explain why the differential equation $\frac{dr}{dt} = \frac{k}{r^2}$ represents the situation.

    (ii) Find the value of k.

    (iii) Hence solve the differential equation to find r in terms of t.

    (iv) Calculate the radius of the oil slick after 1 hour, if left untreated.

    It is planned to treat the oil slick with detergent, and the rate of increase of radius is then expected to be proportional to $\frac{1}{r^2(2+t)}$ with the same initial conditions.

    (b) (i) Write down a differential equation for the new situation involving a constant of proportionality m.

    (ii) Find the value of m.

    (iii) Calculate the new value of the predicted radius after 1 hour.

    Answers:

    (a) (ii) $k = 1620$

    (iii) $r = \sqrt[3]{4860t + 91125}$

    (iv) 72.6 m

    (b) (i) $\frac{dr}{dt} = \frac{m}{r^2(2+t)}$

    (ii) $m = 3240$

    (iii) 49.9 m

4. (a) (i) Use integration by parts to evaluate $\int 4x \cos 2x \, dx$

    (ii) Hence show that $\int 4x \cos^2 x \, dx = x^2 + x \sin 2x + \frac{1}{2} \cos 2x + c$

    Given that $I_n = \int_0^1 x^n e^{-x} dx$, where $n \in \mathbb{Z}$

    (b) (i) Show that $I_0 = k - e^{-1}$ where k is an integer to be evaluated.

    (ii) Integrate by parts to show that $I_n = nI_{n-1} - e^{-1}$ for $n \geq 1$

    (iii) Hence evaluate I_3 leaving your answer in terms of e.

    Answers:

    (a) (i) $2x \sin 2x + \cos 2x + c$

    (b) (i) $k = 1$

    (iii) $6 - 16e^{-1}$

MATHEMATICS: ANALYSIS AND APPROACHES HL

5. A particle moves such that its velocity v ms^{-1} at time t seconds is given by $v(t) = \dfrac{t}{4+t^4}, t \geq 0$.

   (a) (i) Sketch the graph of $y = v(t)$. Indicate the maximum point and write down its coordinates.

   (ii) Use the substitution $u = t^2$ to find $\int \dfrac{t}{4+t^4}\,dt$.

   (iii) Find the exact distance travelled by the particle between $t = 0$ and $t = 3$. Give your answer in the form $k\arctan(h)$, $k, h \in \mathbb{Q}$.

   A second particle has velocity v ms^{-1} such that v is related to the particle's displacement in metres by the function $v(s) = \dfrac{2s}{3+s}$.

   (b) (i) Find the acceleration as a function of s, simplifying your answer.

   (ii) Find the displacement, s, at which the acceleration is 0.14 ms^{-2}. Give your answer to 3SF.

   Answers:

   (a) (i) See sketch:

   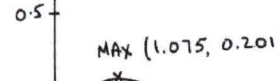
   MAX (1.075, 0.201)

   (ii) $\frac{1}{4}\arctan\left(\frac{1}{2}t^2\right) + c$

   (iii) $\frac{1}{4}\arctan\left(\frac{9}{2}\right)$

   (b) (i) $a(s) = \dfrac{12s}{(3+s)^3}$

   (ii) 0.500 m

6. (a) (i) Use the repeated application of l'Hôpital's rule to find $\lim\limits_{x \to 0} \dfrac{x - \sin x}{x^3}$.

   (ii) Explain why it is valid to apply the rule repeatedly.

   (b) (i) Find the Maclaurin series for $e^{-x}\cos x$ up to the term in x^3.

   (ii) Hence determine the first non-zero term in the Maclaurin series for $e^{-x}\cos x + x - 1$.

   (c) Use the integrating factor method to solve the differential equation $x\dfrac{dy}{dx} + 2y = 10x^2$, given that $y(1) = 3$.

   Answers:

   (a) (i) $\frac{1}{6}$

   (ii) On each application the limit is $\frac{0}{0}$

   (b) (i) $1 - x + \dfrac{x^3}{3}$

   (ii) $\dfrac{x^3}{3}$

   (c) $y = \dfrac{5x^2}{2} + \dfrac{1}{2x^2}$

Chapter 6: MAXIMISING YOUR MARKS

Remember that the examiner is on your side – they want to give you marks! Make it easy for them to follow your thinking, even if you are not quite sure what you are doing or if you are getting wrong answers. You cannot lose marks for doing things wrong. LEARN THIS CHECKLIST.

Before you start a question:

- Read it carefully so you know what it is about.
- Highlight important words.

Answering a question:

- Check any calculations you do, preferably using a different method or order of operation.
- Show your working – there are often marks for method as well as for the right answer. And, in a longer question, a wrong answer at the start may mean lots more wrong answers – but the examiner will probably give you marks for correct methods, and will check your working against your original answer.
- Make sure you have answered *exactly* what the question asked. For example, have you been asked to calculate the new value of an investment or the amount of interest earned.
- In longer questions, don't worry if you can't work out the answer to a part. Carry on with the rest, using their answer (if one is given) or even making up a reasonable answer.
- Don't spend too long on any question or part of a question – you may lose the opportunity to answer easier questions later on. You can always come back and fill in gaps. Work to one mark per minute!
- Use words to explain what you are doing, especially in a longer question.
- The algebra can be tough – keep going!
- Check the units in questions – are they mixed?

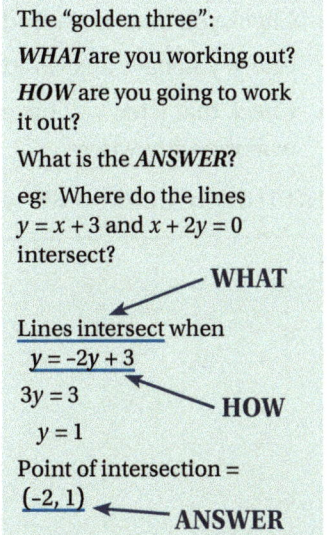

The "golden three":
WHAT are you working out?
HOW are you going to work it out?
What is the *ANSWER*?
eg: Where do the lines $y = x + 3$ and $x + 2y = 0$ intersect?
— WHAT
Lines intersect when
$y = -2y + 3$
$3y = 3$
$y = 1$ — HOW
Point of intersection =
$(-2, 1)$ — ANSWER

MATHEMATICS: ANALYSIS AND APPROACHES HL

Diagrams:

- Do not assume facts from diagrams, especially if they are marked NOT TO SCALE. For example, it may *look* like a right angle but does the question *tell* you that it is? Two lines may *look* parallel but they aren't unless you are *told* they are.
- And do draw your own diagrams – not necessarily to hand in as part of the question, but to help you sort out what's going on.

Key words in questions ("command terms"):

STATE: put the answer down without working (should be an easy one)!

WRITE DOWN: minimal working required.

SHOW: show enough working to get to the given answer.

EVALUATE: give a value to, work out.

SKETCH A GRAPH: draw its shape and show key points (eg: where it cuts the axes)

PLOT A GRAPH: work out points and draw the graph accurately

EXACT VALUE: not a rounded decimal eg: 2π, not 6.28...

> SHOW $x = 3$ is the solution of $2x + 1 = 7$.
> $2 \times 3 + 1 = 7$
> (We have not had to *solve* the equation)

When you have answered the question:

- Check you have answered every part of the question.
- Check you have answered exactly what was asked.
- Check you have answered to the correct accuracy (normally 3SF)
- Check that what you have written is clear, and that your answer is not mixed up in the working somewhere.

DO THESE CHECKS – you will probably pick up a few marks.

Chapter 7: PRACTICE QUESTIONS

The questions which follow are not designed to cover every aspect of the syllabus, nor are they exam style questions. Their purpose is to give you some practice in the **basics**: if you cannot, for example, rearrange an equation with a log function in it, or corectly identify which integral technique to use, then you may be getting questions wrong simply because of a lack of basic techniques. You should answer all of these questions as part of your revision. If you get an answer wrong, find out why: then come back to it later, and see if you can get it right next time.

Number and Algebra

1. Find the 25th term and the sum of the first 54 terms of the sequence which begins: 3, 8, 13, 18...
2. An arithmetic sequence has first term 7 and common difference 3.5. How many terms are required for the sum of the sequence to be 25830?
3. What is the 12th term and the sum to 18 terms of the sequence which begins 3, 12, 48, 192?
4. A geometric series has a first term 400, ten terms and a sum of 1295.67. What is the common ratio?
5. Why does the sum to infinity exist for the sequence 100, 80, 64, 51.2? Find S_{20} and S_∞ and also the percentage error in approximating S_∞ by S_{20}.
6. Write the recurring decimal 0.1343434... as a fraction in its simplest form.
7. Write $2 + 3\log_{10}x$ as a single logarithm.
8. Solve the equation $3.1^x = 10^{x-1}$. Answer to 4 decimal places.
9. If $s = 3 + 10e^{0.4t}$, find t in the form $a \ln b$ when $s = 15$.
10. Solve the equation $4^x - 5 \times 2^x + 4 = 0$ (Hint: Replace 2^x with y)
11. Use your GDC to solve $x + \log_3 x = 10$.
12. A team of 3 is to be chosen from 9 volunteers for a general knowledge contest. How many possible teams are there? If the 9 volunteers consist of 5 boys and 4 girls, how many of the possible teams will have more girls than boys?
13. There are 26 letters in the alphabet and 10 numerical digits. A car has a registration number consisting of 3 letters followed by 2 digits. How many possibilities are there? And how many possibilities if all the letters must be different? If there is a free choice of letters, but the digits cannot begin with a 0, and must form an even number, how many possibilities are there?
14. Find the constant term in the expansion of $\left(3x - \frac{1}{x}\right)^6$.
15. Expand $\sqrt{4 + x}$ up to the term in x^3, stating the values of x for which the expansion is valid.

MATHEMATICS: ANALYSIS AND APPROACHES HL

16. Use mathematical induction to prove that the sum of the first n square numbers is given by the formula $S_n = \frac{1}{6}n(n+1)(2n+1)$.

17. Convert to cis form: $1 + i$, $2 - 3i$, $-6i$, $(2+i)^2$

18. Convert to $a + bi$ form: $2\operatorname{cis} 60°$, $3(\cos\frac{5\pi}{3} + i\sin\frac{5\pi}{3})$, $4e^{i\pi}$

19. Solve $z^2 = 5 - 12i$.

20. Given that $2 + i$ is a root of the equation $z^3 - 11z + 20 = 0$, find the remaining roots.

21. Find the real numbers p for which $1 + pi$ is a solution of $z^2 - 2z + (p+7) = 0$.

22. Find a cubic equation (with real coefficients) which has $3 + i$ and -2 as two of its roots.

23. Work out $(2 - i)^4$ using De Moivre's theorem. Give your answer in both modulus-argument and Cartesian forms. (The argument should be in radians).

24. Find the fifth roots of $1 + 2i$ using De Moivre's theorem.

25. Express as partial fractions: (a) $\dfrac{4x-9}{x^2-5x+6}$ (b) $\dfrac{6x+8}{x^2+4x}$

26. Find the value of a for which the following system of equations does not have a unique solution:
$$\begin{cases} 4x - y + 2z = 1 \\ 2x + 3y = -6 \\ x - 2y + az = k \end{cases}$$

    For this value of a, how many solutions are there if $k = 3.5$, and how many if $k = 0$?

27. Solve the following system of equations:
$$\begin{cases} 3x - 2y + z = -4 \\ x + y - z = -2 \\ 2x + 3y = 4 \end{cases}$$

Functions

1. Find the range of the function $f(x) = \dfrac{x^3-2}{x}$, $x < 0$.

2. Find the largest possible domain of the function $f(x) = \dfrac{1}{\sqrt{9-4x^2}}$.

3. Why is the inverse of $f(x) = x(x-2)$, $x \in \mathbb{R}$ not a function? Suggest a domain restriction which would ensure that $f^{-1}(x)$ *is* a function.

4. If $f(x) = x + 1$ and $g(x) = x^3$ find the function $(f \circ g)^{-1}$.

5. Given $f(x) = 2x + 1$ and $g(x) = \cos x$, $0 \leq x \leq \pi$, solve the equation $(g \circ f)(x) = 0.8$.

6. State whether each of the following functions is odd, even, or neither:
$$f(x) = 4 - 3x^2,\ f(x) = 3x^2 + x,\ f(x) = x - \tfrac{1}{x},\ f(x) = x^2 + |x|,\ f(x) = x^3 + |x|$$

7. For the graph of $f(x) = \dfrac{e^{-x}}{(x+1)^2}$, $x \neq -1$, identify any horizontal and vertical asymptotes. Find the turning point, and the solutions of the equation $f(x) = 7$.

8. Define the transformations which will transform the graph of $y = \sqrt{x}$ into the graph of $y = 3\sqrt{a-x}$, where a is a constant. Sketch the two graphs on the same axes.

9. The graph of $f(x)$ (see sketch) has a horizontal asymptote at $y = -2$ and a vertical asymptote at $x = 1$. It passes through the origin. Sketch the graphs (on separate axes) of $y = \dfrac{1}{f(x)}$, $y = f^{-1}(x)$, $y = |f(x)|$, $y = f(|x|)$, $y = f(|x|)$.

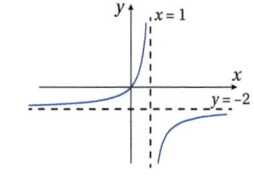

7. PRACTICE QUESTIONS

10. Use the quadratic formula to solve $x + 3 = \frac{2}{x}$, giving your answer in exact form.
11. Complete the square for: $x^2 - 4x + 2$, $2x^2 + 6x + 5$, $12 - 2x - x^2$.
12. For each of the quadratics in 11, write down the turning point and the line of symmetry.
13. Find the range of values of k for which $2x^2 + 2x + k = 0$ has two real, distinct solutions.
14. Use factorisation to find all the real solutions to $x^4 + 2x^2 - 15 = 0$.
15. Find the roots of $x^2 - 6x + k = 0$ given that one root is three times the other and k is a constant. Also find the value of k. *(Hint: Use sum and product of roots for α and 3α)*.
16. Work out the inverses of these functions: $f(x) = 4 \times 3^x$; $f(x) = \frac{1}{\sqrt{\ln x + 4}}$; $f(x) = \frac{x}{x-1}, x \neq 1$
17. Solve the inequality $x^2 - 7x + 6 \leq 0$. Hence solve $(e^x - 2)(e^x - 3) < 2e^x$.
18. Use your GDC to solve the inequality $|4x - 1| < 5$
19. Use your GDC to find the values of x for which $|5 - 3x| \leq |x + 1|$.
20. The graph of a cubic function cuts the x-axis at $(-2, 0)$ and has a maximum at $(4, 0)$. The y-intercept is at $(0, -16)$. Write down its equation.
21. Prove that $(x + 2)$ is a factor of $x^4 - 3x^2 + 3x + 2$.
22. The polynomial $x^3 + ax^2 + bx + 5$ leaves a remainder of 2 when divided by $(x - 1)$ and a remainder of 9 when divided by $(x - 2)$. Find the values of a and b.
23. The graphs $y = x^3 - x^2 - x + 4$ and $y = x + 4$ intersect at A, B and C, where B is between A and C. Find the distance AC.
24. Solve the equation $xe^x = 4 - 2x$, giving your answer to 5DP.

Geometry and Trigonometry

1. Convert to radians, giving answers in an exact form: 30°, 45°, 120°, 330°.
2. The sector of a circle with radius 5 cm has an arc length of 12 cm. Find the angle of the sector in radians, and its area.
3. Solve the equation $\sec^2 \theta = 3$, $0° \leq \theta \leq 360°$.
4. If $\sin \theta = \frac{3}{8}$ and θ is obtuse, find the exact values of $\cos \theta$ and $\tan \theta$.
5. Using the data in question 4, work out the exact values of $\sin 2\theta$, $\cos 2\theta$ and $\tan 2\theta$.
6. Find an identity for $\sin 3\theta$ in terms of $\sin \theta$. (You will need to use the compound angle identities, the double angle identities and the Pythagorean identities).
7. Write down the domain and range of each of the functions $f(x) = \cos x$ and $f(x) = \cos^{-1} x$.
8. Write down the equation of the function shown in the diagram in the form $y = a \sin(b(x + c))° + d$

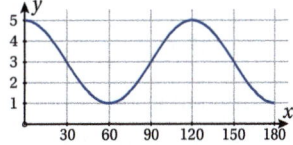

Use the trigonometric identities in questions 9–12.

9. Solve $2\sin x = 5\cos x$, $0 \leq x \leq 2\pi$
10. Solve $2\sin x = \cos 2x$, $-180° \leq x \leq 180°$
11. Solve $2\sin 2\theta = 3\sin \theta$, $0 \leq \theta \leq \pi$
12. Solve $3\sin x = \tan x$, $0 \leq x \leq 360°$
13. Solve the following triangles (the triangle in each case is ABC):

    BC = 6 cm, C = 87°, A = 45°. Find AB.

AB = 6 cm, A = 87°, AC = 5.4 cm. Find BC.

AB = 6 cm, BC = 5.4 cm, CA = 3.5 cm. Find B.

AB = BC = 5.2 cm. B = 34°. Find AC.

AC = 6 cm, C = 32°, A = 90°. Find AB.

BC = 6 cm, AB = 4 cm, C = 25°. Find A. (Two possibilities).

14. Find the area of the first and second triangles in question 13.

15. A cylindrical tin has a closed base but is open at the top. Find its surface area and volume if $r = 3.5$ and $h = 6$.

16. A mast OT has is stabilised by two guy ropes, TA and TB, with angles of depression 65° and 20° respectively. OAB is a straight line. Given that the mast is 15 m tall, find the length AB, and the total length of rope required.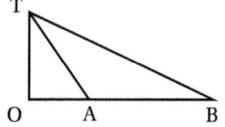

17. A square base pyramid has its apex directly above the centre of the base. If the base edge is 6 cm and the height of the pyramid is 9 cm, find the length of a slant edge, and the angle a slant edge makes with the base.

18. Solve $\sin 2x = \sqrt{3} \cos x$, $0 \leq x \leq \pi$, giving your answers in terms of π.

19. Express $\dfrac{\sqrt{3} + \tan x}{1 - \sqrt{3} \tan x}$ (x in radians) as a single trigonometric ratio.

20. The sides of a triangle are x, $x + 1$ and p, where $p > x + 1$. If the largest angle is 120°, find an expression for p in terms of x. Find x if $p = \sqrt{7}$.

21. If A = (1, 4), B = (3, -2), C = (-1, -4) and D = (3, 5), find vectors $\overrightarrow{AB}, \overrightarrow{BC}, \overrightarrow{CD}, \overrightarrow{AD}, \overrightarrow{BD}$. Which two vectors are parallel?

22. Find p and q such that $(p\mathbf{i} + p\mathbf{j}) + (3\mathbf{i} + 2q\mathbf{j}) = (q\mathbf{i} + 18\mathbf{j})$.

23. If $\mathbf{r} = 2\mathbf{i} + 3\mathbf{j} - \sqrt{3}\mathbf{k}$, find the magnitude of $\mathbf{r}$ and the unit vector in the direction of $\mathbf{r}$.

24. Find the value of a such that vectors $3\mathbf{i} + 6\mathbf{k}$ and $2\mathbf{i} + \mathbf{j} + a\mathbf{k}$ are perpendicular.

25. If $\mathbf{a}$, $\mathbf{b}$ and $\mathbf{c}$ are three non-zero vectors, which of the following statements must be true:

   (i) $\mathbf{a} \cdot \mathbf{b} = \mathbf{a} \cdot \mathbf{c} \Rightarrow \mathbf{b} = \mathbf{c}$ (ii) $\mathbf{a} \cdot \mathbf{b} = 0 \Rightarrow \mathbf{a} \perp \mathbf{b}$ (iii) $\mathbf{a} \cdot (\mathbf{b} + \mathbf{c}) = \mathbf{a} \cdot \mathbf{b} + \mathbf{a} \cdot \mathbf{c}$
   (iv) $\mathbf{a} \cdot (\mathbf{b} \cdot \mathbf{c}) = (\mathbf{a} \cdot \mathbf{b}) \cdot \mathbf{c}$

26. Find the angle between the vectors $2\mathbf{i} - 5\mathbf{j} + \mathbf{k}$ and $4\mathbf{i} + 2\mathbf{j} + 3\mathbf{k}$.

27. A = (1, 2, 0), B = (1, 4, -3), C = (6, -2, 4), D = (0, 3, 3). Use the scalar product to find the angle between lines AB and CD.

28. Write down the equation of the line AB (points as in number 27) in both vector and Cartesian form.

29. Find the point where the lines $\mathbf{i} + \mathbf{j} - \mathbf{k} + \lambda(2\mathbf{i} + \mathbf{j} + 2\mathbf{k})$ and $\dfrac{x-2}{2} = \dfrac{9-y}{2} = \dfrac{z}{2}$ meet.

30. Prove that the lines $\mathbf{i} + 2\mathbf{j} + \mathbf{k} + \lambda(-\mathbf{j} + 4\mathbf{k})$ and $4\mathbf{k} + \mu(-\mathbf{i} + \mathbf{j} - 5\mathbf{k})$ are skew.

31. How can you tell that two lines are parallel from their vector equations?

32. Find a vector which is perpendicular to both the lines in number 27.

33. If O is the origin, P = (1, 2, 1) and R = (-1, 4, 0), find point Q such that OPQR is a parallelogram. Use the cross product to find the area of the parallelogram.

34. Find the Cartesian equation of the plane which is parallel to the parallelogram in number 33, and which contains the point (3, 1, -2).

7. PRACTICE QUESTIONS

35. A = (1, 0, 1), B = (2, 2, 1), C = (3, -1, 0). Find the cross product $\vec{AB} \times \vec{BC}$ and hence the Cartesian equation of plane ABC.

36. Find where the line $r = 4i + j + 2k + \lambda(-2i + j)$ meets the plane in question 35. Find also the angle between the line and the plane.

37. Find the point where the three planes $2x + 4y + z = 11$, $2x - y + 3z = 5$ and $x + z = 3$ intersect.

38. The plane $3x - 2y + 2z = 5$ contains the line $\frac{x-a}{2} = y - 1 = \frac{z+1}{b}$. Find the values of a and b.

39. A plane has vector equation $\begin{pmatrix} x \\ y \\ z \end{pmatrix} = \begin{pmatrix} 4 \\ -1 \\ 0 \end{pmatrix} + \lambda \begin{pmatrix} 1 \\ 1 \\ 2 \end{pmatrix} + \mu \begin{pmatrix} 0 \\ 2 \\ -1 \end{pmatrix}$. Find its Cartesian equation.

Statistics and Probability

The amount spent (in €) by the first 50 people going into a shop is shown in the table below:

15.60	5.95	31.22	3.02	6.60	24.70	15.45	32.50	12.45	4.43
12.65	10.09	52.86	12.88	2.53	31.79	9.86	25.79	18.28	32.05
14.87	24.65	15.70	8.65	4.42	17.20	8.53	0.45	0.95	4.44
7.45	5.82	45.20	2.70	10.04	15.70	32.20	12.43	36.75	32.50
16.87	3.78	0.56	33.67	9.67	25.50	33.06	7.56	2.63	45.80

Questions 1 to 9 refer to this table.

1. Is this data discrete or continuous?
2. Draw up a grouped frequency table (with first group €0.01 - €10.00). You should have 6 groups.
3. Which is the modal group?
4. Enter the mid-values of each group and the frequencies onto your GDC. Calculate estimates of the mean and the standard deviation. (Why "estimates"?)
5. Draw a frequency histogram to represent the data.
6. Complete a cumulative frequency table for the data, and hence draw a cumulative frequency graph.
7. From the cumulative frequency graph, write down the median, the lower quartile, the upper quartile and calculate the interquartile range.
8. Draw a box and whisker plot for the data. Is the maximum value an outlier?
9. What was the least amount that the people in the top ten percentiles spent?
10. Find the correlation coefficient and the regression line of y on x for the following data:

x	2.0	2.5	3.0	3.5	4.0	4.5	5.0
y	12.56	28.23	15.10	45.54	30.35	35.76	42.19

11. A discrete probability distribution has $E(X) = 14.8$ and $Var(X) = 4.9$. Calculate the expected mean and the expected standard deviation of the distribution $3X - 4$.

12. I have 6 red socks and 4 green socks in a drawer. I take 2 out at random. Draw a tree diagram to show the possible outcomes and find P(the two socks do not match).

13. A and B are two events such that P(A) = 0.2, P(B) = 0.5 and P(A ∪ B) = 0.55. Use a Venn diagram to find: P(A ∩ B); P(A' ∩ B); P(A|B); P(B'|A).

14. Two dice are rolled. Find the probability that they show different numbers given that the total is 8.

15. Given that P(A ∪ B) = 0.7, P(A) = 0.6 and that A and B are independent events, find P(B).

16. Given that $P(A) = \frac{2}{3}$, $P(B|A) = \frac{2}{5}$ and $P(B|A') = \frac{1}{4}$, find P(B') and P(A' ∪ B').

17. The probability of contracting a serious disease after being bitten by a particular venomous snake is 5 in 1000. A test carried out on someone who has been bitten by a snake will correctly diagnose this disease in 90% of cases, but will *incorrectly* diagnose the disease in 12% of cases. Find the following probabilities:
    (a) That the test gives a positive result.
    (b) That the test gives a correct diagnosis.
    (c) That given a positive test, the patient does *not* have the disease.

18. The probability distribution for a discrete random variable X is as follows:

x	1	2	3	4	5
P(X = x)	0.3	0.35	k	$2k$	0.05

    Find the value of k and the expected mean.

19. A probability density function which models the life expectancy of new-born infants is given by: $f(x) = \begin{cases} kx^2(90-x), & \text{for } 0 < x < 90 \\ 0 & \text{otherwise} \end{cases}$
    Find the value of k; the expected mean; the commonest age at which people would be expected to die; the median; and the proportion of the population who would be expected to live beyond 80.

20. For $X \sim B(12, 0.2)$, find $P(X = 3)$, $P(X \leq 2)$, $P(X > 4)$. Find also the mean and the variance.

21. For $X \sim B(6, p)$, $P(X = 5) = 0.393216$. Find p.

22. For $X \sim B(4, 0.2)$, find $P(X = 3)$, without using your calculator. *(Hint: you may find this easier if you work in fractions.)*

23. If $X \sim N(100, 5^2)$, find $P(X < 112)$, $P(X < 91)$, $P(95 < X < 101)$.

24. X is a normally distributed variable with $\mu = 18$. If $P(X > 20) = 0.115965$, find the standard deviation.

25. For a certain type of potato, those with weight less than 100 g are branded "small", and those with a weight over 125 g are branded "large". In one batch, 2.55% of potatoes are small, 13.61% are large. Assuming the weights are normally distributed, calculate the mean and standard deviation of the batch.

Calculus

1. Differentiate $f(x) = x^3$ and $f(x) = 2x^2 - 3x$ from first principles.

2. Differentiate these functions:
   (a) xe^{-x} (b) $3\arcsin x$ (c) $\cos^2 2x$ (d) $4\sqrt{x} - 5$ (e) $2\ln(\cos x)$ (f) $\frac{x^2-2}{x}$ (g) $\frac{3x^3}{(x+1)}$
   (h) $\sqrt{x^3-2}$ (i) $\sec^2 x$ (j) 3×4^x (k) $x^2 \log_3 x$ (l) $\frac{x^2}{\tan x}$ (m) $\ln(3-x^2)$

3. If $y = \arccos(1 - 2x^2)$, find $\frac{dy}{dx}$, simplifying your answer.

7. PRACTICE QUESTIONS

4. Given that $y = \frac{x^2-1}{2x^2+1}$, find the set of values of x for which $\frac{dy}{dx} > 0$. Find the coordinates of any stationary points.

5. Find the equations of the tangent and normal to $y = 3\ln x$ at the point with x-coordinate 3.

6. Find the equations of the tangents to the curve $y^2 + 3xy + 4x^2 = 14$ at the points where $x = 1$.

7. Find the first and second derivatives of $y = xe^x$.

8. Find the coordinates of all stationary points and the point of inflexion on the graph of the function $f(x) = x^3 - 3x^2 + 1$. What is the gradient at the point of inflexion? (No GDC).

9. For the graph of the function $f(x) = \frac{x^2-1}{x}$ find: any axis intercepts; the vertical asymptote; any turning points. (No GDC).

10. A circular oil slick is increasing in radius at the rate of 2 m/min. Find, in terms of π, the rate at which the area of the slick is increasing when its radius is 30 m.

11. Use l'Hôpital's rule to find $\lim_{x \to 1} \frac{x-1}{\sqrt{x}-1}$.

12. Integrate these functions: (a) $\int \sin 3x \, dx$ (b) $\int \frac{4x}{x^2-1} dx$ (c) $\int \frac{1}{4+x^2} dx$ (d) $\int x\sqrt{2x-3} \, dx$ (e) $\int \frac{1+x}{\sqrt{1-x^2}} dx$ (f) $\int 3\cos^2 x \, dx$ (g) $\int xe^{2x} dx$ (h) $\int \frac{1}{(2-x)^2} dx$ (i) $\int e^x \sin x \, dx$ (j) $\int \frac{2x+1}{x^2-4} dx$

13. Find the real number $k > 1$ for which $\int_1^k \left(1 + \frac{1}{x^2}\right) dx = \frac{3}{2}$

14. Find the area enclosed by the curve $y = 4x - x^2$ and the x-axis.

15. Find the area enclosed between the graph of $y = x\cos(x^2)$, the x-axis, $x = 0$ and the next positive x-intercept.

16. Find the area enclosed between the curves $y = 2x^2 + 3$ and $y = 10x - x^2$.

17. Find the volume enclosed when the area lying in the first quadrant and bounded by the curve $y = 2x^2 + 1$ between $y = 2$ and $y = 4$ is rotated 360° around the y-axis.

18. Find $f(x)$ given that $f'(x) = 2x(5-x)$ and that $y = 3$ when $x = 0$.

19. Find the distance travelled by a particle between $t = 0$ and $t = 5$ given that $v = 8 - 2t$.

20. The displacement s of a particle from an origin O at time t seconds is $s = 2t^2 - 3t + 6$. Find the displacement, the velocity and the acceleration of the particle when $t = 1.5$.

21. A particle moves in a straight line. At time t secs its acceleration is given by $a = 3t - 1$. When $t = 0$, the velocity of the particle is $2\,\text{ms}^{-1}$ and it is 3 m from the origin. Find expressions for v and s in terms of t. Show that the particle is always moving away from the origin.

22. Differentiate $\frac{2-3x}{x^2+3x+3}$ and hence find the turning points on the graph of $y = \frac{2-3x}{x^2+3x+3}$. Sketch the graph.

23. Find the particular solutions of the following differential equations:

    (a) $(x^2-1)\frac{dy}{dx} = y$, $y = \sqrt{2}$ when $x = 3$;

    (b) $x^2 \frac{dy}{dx} = y + 3$, $y = -2$ when $x = 1$;

    (c) $\frac{dy}{dx} = \sqrt{1-y^2}$, $y = 0$ when $x = \frac{\pi}{6}$.

24. Use Euler's method to find y when $x = 1$, given that $\frac{dy}{dx} = x^2 + y^2$, and that $y = 0$ when $x = 0.5$. Use a step length of $x = 0.1$.

MATHEMATICS: ANALYSIS AND APPROACHES HL

25. Why is the differential equation $x^2 \dfrac{dy}{dx} = 1 + xy$ not homogeneous?

26. Find the general solution of $xy\dfrac{dy}{dx} = x^2 + y^2$ using the substitution $y = vx$.

27. Use the integrating factor to find the general solution of $\dfrac{dy}{dx} + 3y = e^{2x}$.

28. Find the Maclaurin series for $\arctan x$ as far as the term in x^3, and find the percentage error when using the series to calculate $\arctan 0.5$.

Answers to Practice Questions

Number and Algebra

1. 123, 7317
2. 120
3. 12 582 912, 6.87×10^{10}
4. 0.7
5. $r = 0.8$, 494.24, 500, 1.15%
6. 133/990
7. $\log_{10} 100x^3$
8. 1.9660
9. $2.5 \ln 1.2$
10. 0 or 2
11. 8.096
12. 84, 34
13. 1 757 600, 1 560 000, 7 909 200
14. −540
15. $2 + \tfrac{1}{4}x - \tfrac{1}{64}x^2 + \tfrac{1}{512}x^3, |x| < 4$

16. $\tfrac{1}{6}n(n+1)(2n+1) + (n+1)^2$
    $= \tfrac{1}{6}(n+1)(n+2)(2n+3)$; $n = 1$ gives 1.
17. $\sqrt{2}\operatorname{cis} 45°$, $\sqrt{13}\operatorname{cis} 56°$, $6\operatorname{cis}(-90°)$, $5\operatorname{cis} 53.1°$
18. $1 + i\sqrt{3}$, $\tfrac{3}{2} - \tfrac{3\sqrt{3}}{2}i$, −4
19. $\pm(3 - 2i)$
20. $2 - i$, −4
21. −2 or 3
22. $z^3 - 4z^2 - 2z + 20$
23. $-7 - 24i$, $25\operatorname{cis} 1.855$
24. $1.17\operatorname{cis}(12.68 + 72n)°$, $n = 0, 1, 2, 3, 4$
25. $\dfrac{3}{x-3} + \dfrac{1}{x-2}$, $\dfrac{2}{x} + \dfrac{4}{x+4}$
26. $a = 1; \infty, 0$
27. $x = -1, y = 2, z = 3$

Functions

1. $f(x) \geq 3$
2. $-1.5 < x < 1.5$
3. It's 1-many; $x \geq 1$ (others possible)
4. $(f \circ g)^{-1} = \sqrt[3]{(x-1)}$
5. 2.32
6. E, N, O, E, N

7. $y = 0$, $x = -1$, $(-3, 5.02)$, $x = -4.38$ or -0.512 or -2.06
8. Reflect in y-axis, translate $\begin{pmatrix} a \\ 0 \end{pmatrix}$, stretch ×3 in y-direction.

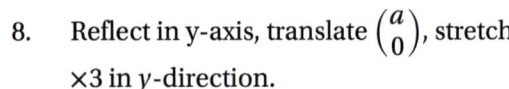
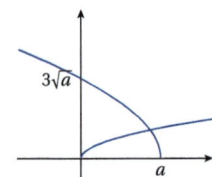

9.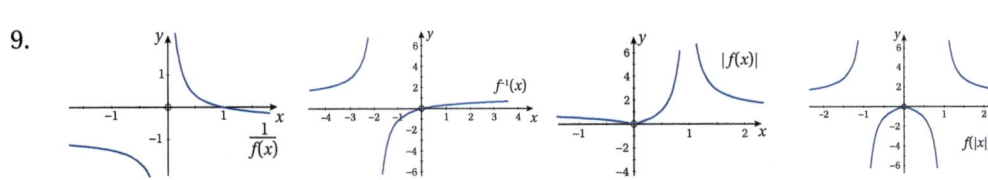

7. PRACTICE QUESTIONS

10. $\frac{-3 \pm \sqrt{17}}{2}$
11. $(x-2)^2 - 2$, $2(x+1.5)^2 + 0.5$, $13 - (x+1)^2$
12. $(2, -2)$, $x = 2$; $(-1.5, 0.5)$, $x = -1.5$; $(-1, 13)$, $x = -1$
13. $k < 0.5$
14. $\pm\sqrt{3}$
15. 1.5, 4.5; 6.75
16. $\log_3 \frac{x}{4}$, $e^{(\frac{1}{x^2} - 4)}$, $\frac{x}{x-1}$

17. $1 < x < 6$, $0 < x < \ln 6$
18. $-1 < x < 1.5$
19. $1 \leq x \leq 3$
20. $y = 0.5(x+2)(x-4)^2$
21. $(-2)^4 - 3(-2)^2 + 3(-2) + 2 = 0$
22. $a = 2$, $b = -6$
23. $\sqrt{18}$
24. 0.88652

Geometry and Trigonometry

1. $\frac{\pi}{6}, \frac{\pi}{4}, \frac{2\pi}{3}, \frac{11\pi}{6}$
2. 2.4, 30
3. 54.7°, 125.3°, 234.7°, 305.3°
4. $-\frac{\sqrt{55}}{8}, -\frac{3}{\sqrt{55}}$
5. $-\frac{3\sqrt{55}}{32}, \frac{23}{32}, -\frac{3\sqrt{55}}{23}$
6. $3\sin\theta - 4\sin^3\theta$
7. All reals, $-1 \leq \cos x \leq 1$; $-1 \leq x \leq 1$, $0 \leq \cos^{-1} x \leq \pi$
8. $y = 2\sin(3(x+30)) + 3$
9. 1.19, 4.33
10. 21.5°, 158.5°
11. $0, \pi, 0.723$
12. 0°, 70.5°, 180°, 289.5°, 360°
13. 8.47, 7.86, 35.3°, 3.04, 3.75, 39.3° or 140.7°
14. 18.9 cm², 16.2 cm²
15. 170.4, 230.9
16. 34.2 m, 60.4 m
17. 9.95 cm, 64.8°
18. $\frac{\pi}{2}, \frac{\pi}{3}, \frac{2\pi}{3}$
19. $\tan(\frac{\pi}{3} + x)$
20. $p = \sqrt{3x^2 + 3x + 1}$, $x = 1$.

21. $\begin{pmatrix} 2 \\ -6 \end{pmatrix}, \begin{pmatrix} -4 \\ -2 \end{pmatrix}, \begin{pmatrix} 4 \\ 9 \end{pmatrix}, \begin{pmatrix} 2 \\ 1 \end{pmatrix}, \begin{pmatrix} 0 \\ 7 \end{pmatrix}$ BC, AD
22. $p = 4$, $q = 7$
23. 4, $r = \frac{1}{2}i + \frac{3}{4}j - \frac{\sqrt{3}}{4}k$
24. -1
25. (ii), (iii)
26. 88.1°
27. 62.7°
28. $r = i + 2j + \lambda(2j - 3k)$, $x = 1$, $\frac{y-2}{2} = \frac{z}{-3}$
29. (7, 4, 5)
30. To equate i and j, $\mu = -1$, $\lambda = 3$. This gives $14k$ and $9k$, so do not intersect. Also not parallel, so skew.
31. Same direction vectors.
32. $13i + 18j + 12k$
33. $(0, 6, 1)$, $\sqrt{53}$
34. $4x + y - 6z = 25$
35. $2i - j + 5k = 7$, $2x - y + 5z = 7$
36. $(0, 3, 2)$, 24.1°
37. $(2\frac{2}{3}, 1\frac{1}{3}, \frac{1}{3})$
38. 3, -2
39. $5x - y - 2z = 21$

189

MATHEMATICS: ANALYSIS AND APPROACHES HL

Statistics and Probability

1. Discrete.

2.
0.01 – 10.00	10.01 – 20.00	20.01 – 30.00	30.01 – 40.00	40.01 – 50.00	50.01 – 60.00
20	14	5	8	2	1

3. 0.01 – 10.00

4. 17.2, 13.31; not using original data.

5.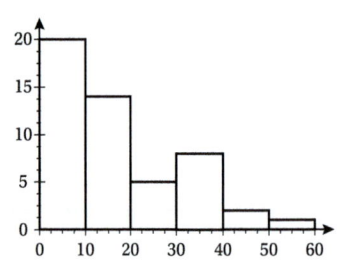

6.
€	≤10	≤20	≤30	≤40	≤50	≤60
c.f.	20	34	39	47	49	50

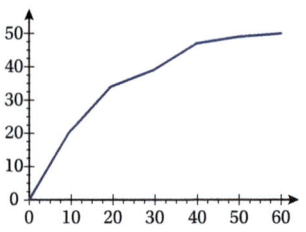

7. $Q_1 = 6$, $Q_2 = 14$, $Q_3 = 28$, IQR = 22

   No. $Q_3 + 1.5 \times$ IQR = 61

8.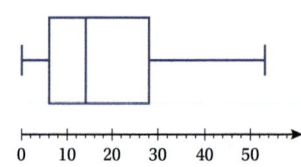

9. €38

10. 0.730, $y = 8.51x + 0.161$

11. 40.4, 6.64

12. P(no match) = 24/45; see diagram:

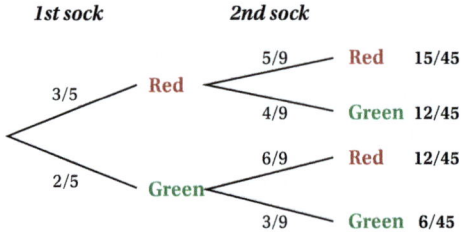

13. 0.15, 0.35, 0.3, 0.25

14. 0.8

15. 0.25

16. $\frac{13}{20}, \frac{11}{15}$

17. 0.124, 0.880, 0.964

18. 0.1, 2.35, 1.5275

19. 1.829×10^{-7}, 54, 60, 55.3, 6.3%

20. 0.236, 0.558, 0.0726; 2.4, 1.92

21. $\frac{256}{10000} = 0.0256$

22. 0.8

23. 0.992, 0.0359, 0.421

24. 3.5

25. 116, 8.2

7. PRACTICE QUESTIONS

Calculus

1. $\lim_{x \to 0} \dfrac{(x+h)^3 - x^3}{h} = \lim_{x \to 0} \dfrac{3x^2 h + 3h^2 x + h^3}{h} = 3x^2$,

   $\lim_{x \to 0} \dfrac{(2(x+h)^2 - 3(x+h)) - (2x^2 - 3x)}{h} = \lim_{x \to 0} \dfrac{4xh + 2h^2 - 3h}{h} = 4x - 3$

2. (a) $e^{-x}(1-x)$ (b) $\dfrac{3}{\sqrt{1-x^2}}$ (c) $-4\sin 2x \cos 2x$ (d) $\dfrac{2}{\sqrt{x}}$

   (e) $-2\tan x$ (f) $1 + \dfrac{2}{x^2}$ (g) $\dfrac{3x^2(2x+3)}{(x+1)^2}$ (h) $\dfrac{3x^2}{2\sqrt{x^3-2}}$

   (i) $2\sec^2 x \tan x$ (j) $3\ln 4 \times 4^x$ (k) $\dfrac{x}{\ln 3} + 2x\log_3 x$ (l) $\dfrac{2x\tan x - x^2 \sec^2 x}{\tan^2 x}$

   (m) $-\dfrac{2x}{3-x^2}$

3. $\dfrac{2}{\sqrt{1-x^2}}$

4. $x > 0$, $(0, -1)$

5. $y = x + 0.296$, $y = 6.296 - x$

6. $x + y + 4 = 0$, $y + 2x = 4$

7. $e^x(x+1)$, $e^x(x+2)$

8. $(0, 1)$, $(2, -3)$; $(1, -1)$; -3

9. $(-1, 0)$, $(1, 0)$; $x = 0$; none

10. 120π

11. 2

12. (a) $-\dfrac{1}{3}\cos 3x + c$ (b) $2\ln(x^2 - 1) + c$ (c) $\dfrac{1}{2}\arctan\dfrac{x}{2} + c$

   (d) $\dfrac{1}{10}(2x-3)^{\frac{5}{2}} + \dfrac{1}{2}(2x-3)^{\frac{3}{2}} + c$ (e) $\arcsin x - \sqrt{1-x^2} + c$ (f) $\dfrac{3}{2}x + \dfrac{3}{4}\sin 2x + c$

   (g) $\dfrac{x}{2}e^{2x} - \dfrac{1}{4}e^{2x} + c$ (h) $\dfrac{1}{2-x} + c$ (i) $\dfrac{1}{2}e^x(\sin x - \cos x) + c$

   (j) $\dfrac{5}{4}\ln(x-2) + \dfrac{3}{4}\ln(x+2) + c$

13. 2

14. $10\dfrac{2}{3}$

15. 0.5

16. $9\dfrac{13}{27}$

17. 2π

18. $y = 5x^2 - \dfrac{2}{3}x^3 + 3$

19. 17

20. $6, 3, 4$

21. $v = 1.5t^2 - t + 2$, $s = 0.5t^3 - 0.5t^2 + 2t + 3$

    $v \neq 0$ for any value of t (discriminant < 0). So v is always positive, and particle is moving away from the origin.

22. $\dfrac{3x^2 - 4x - 15}{(x^2 + 3x + 3)^2}$, $(3, -\dfrac{1}{3})$, $(-\dfrac{5}{3}, 9)$

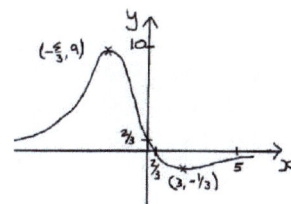

23. $y = 2\sqrt{\dfrac{x-1}{x+1}}$, $y = e^{1-\frac{1}{x}} - 3$, $y = \sin\left(x - \dfrac{\pi}{6}\right)$.

24. 0.2597

25. $\dfrac{dy}{dx} = \dfrac{1}{x^2} + \dfrac{y}{x}$ which is not in the form $f\left(\dfrac{y}{x}\right)$

26. $y^2 = x^2 \ln(Ax^2)$

27. $y = \dfrac{1}{5}e^{2x} + Ae^{-3x}$

28. $x - \dfrac{x^3}{3}$; 1.15%